WEIRD DINOSAURS

WEIRD DINOSAURS

THE STRANGE NEW FOSSILS CHALLENGING EVERYTHING WE THOUGHT WE KNEW

JOHN PICKRELL

Columbia University Press
New York

Columbia University Press
Publishers Since 1893
New York   Chichester, West Sussex
cup.columbia.edu

First published in Austrailia by NewSouth, an imprint of UNSW Press Ltd.
Copyright © 2016 John Pickrell
All rights reserved

Library of Congress Cataloging-in-Publication Data

Names: Pickrell, John.
Title: Weird dinosaurs / John Pickrell.
Description: New York : Columbia University Press, [2017] | "Originally published in Australia by University of New South Wales Press, 2016." | Includes bibliographical references and index.
Identifiers: LCCN 2016057392 (print) | LCCN 2016059780 (ebook) | ISBN 9780231180986 (cloth : alk. paper) | ISBN 9780231543392 (e-book : alk. paper) | ISBN 9780231543392 (e-book)
Subjects: LCSH: Dinosaurs. | Animals, Fossil. | Paleontology. | Paleontological excavations.
Classification: LCC QE861.4 .P53 2017 (print) | LCC QE861.4 (ebook) | DDC 567.9--dc23
LC record available at https://lccn.loc.gov/2016057392

Columbia University Press books are printed on permanent and durable acid-free paper.
Printed in the United States of America

Book design Jo Pajor-Markus
Cover design Xou Creative
Cover images FRONT: Spinosaurus, © Davide Bonadonna BACK: *Top* Yi qi, © Emily Willoughby *Bottom* Regaliceratops, © Royal Tyrrell Museum of Paleontology/ Julius Csotonyi

For my mother, who encouraged my passion for natural history.

CONTENTS

	World map	x
	Foreword by Philip Currie	xiii
	Introduction: A new golden age for dinosaur science	1
1	Monster from the Cretaceous lagoon	9
	The Sahara, Egypt	
2	All hail the dino-bat	27
	Hebei Province, China	
3	Dwarf dinosaurs and trailblazing aristocrats	47
	Transylvania, Romania	
4	Horny ornaments and sexy ceratopsians	65
	Alberta, Canada	
5	The 'unusual terrible hands'	80
	Gobi Desert, Mongolia	
6	Scandalous behaviour and enfluffled vegetarians	98
	Siberia, Russia	
7	Cretaceous creatures of the frozen north	115
	Alaska, United States	

Contents

8	The hidden treasures Down Under	135
	Lightning Ridge, Australia	
9	Record-breaking titans	153
	Patagonia, Argentina	
10	Southern killers set adrift	170
	Mahajanga Basin, Madagascar	
11	Polar pioneers and the frozen crested lizard	190
	Transantarctic Mountains, Antarctica	
	Future potential	212
	Glossary	216
	Further reading	219
	Acknowledgments	220
	Notes	221
	Credits	230
	Index	231

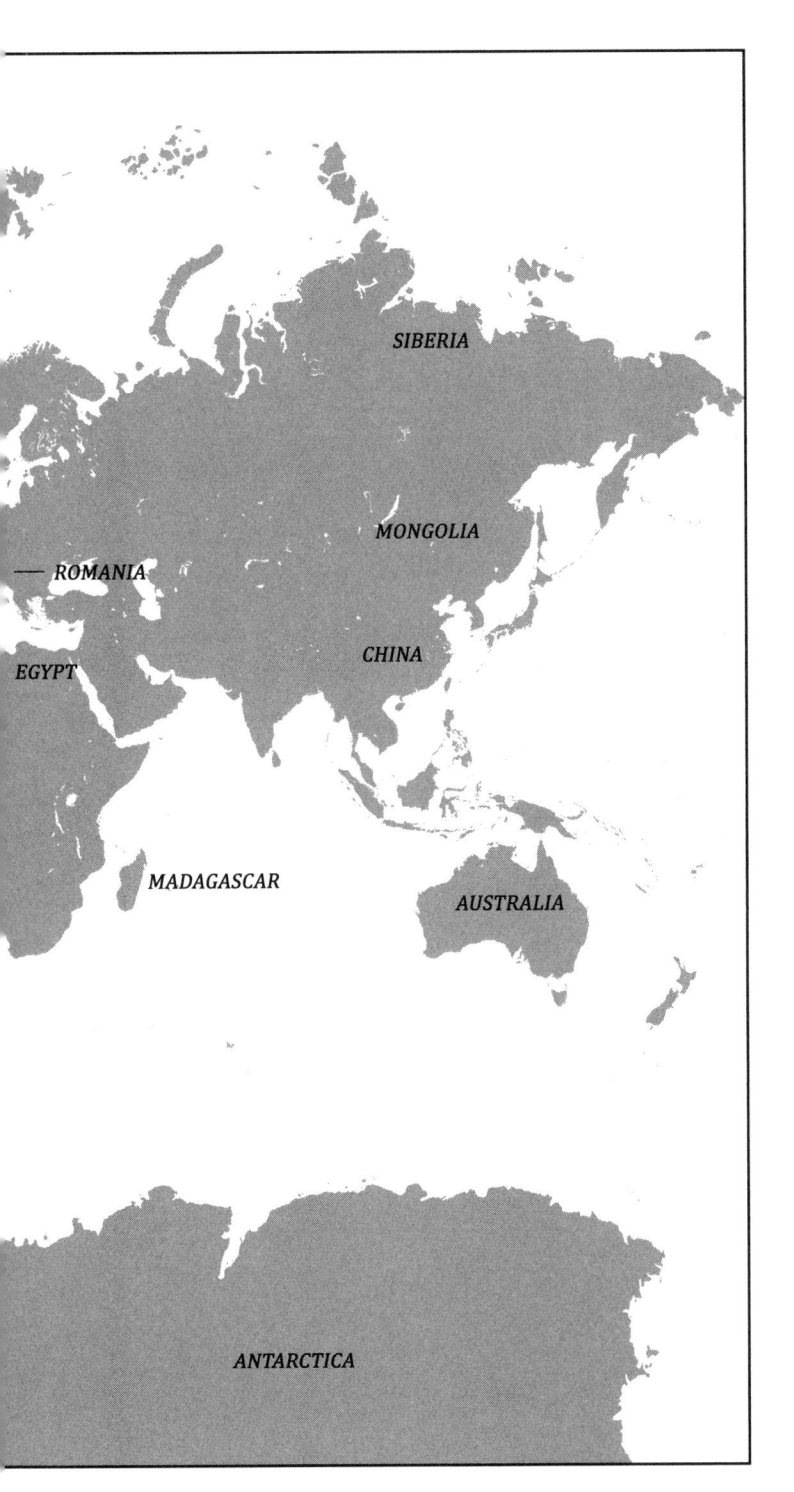

FOREWORD

As a kid growing up in eastern Canada in the 1950s, I was proud of the fact that I knew every dinosaur name that I ever found in a book. And I thought that after more than a century of research on dinosaurs we must have known most of the dinosaurs that had ever lived. Fortunately for my future career choice, I was proved wrong about that.

Today, I am amazed that there are still children who can recite great lists of dinosaur names – even new names with complex pronunciations from China and other countries. The difference is that today they have perhaps ten times as many names to learn, and the rules for naming them are much less straightforward. Furthermore, we now realise that our knowledge of dinosaur diversity is but a small fraction of what it must have been during the Mesozoic Era.

As an example, one of the richest known dinosaur sites in the world is Dinosaur Provincial Park in Alberta, Canada. The rocks are Cretaceous in age, and represent a coastal floodplain spanning a narrow window of time that covers less than 5 million years. More than 45 dinosaur species have been identified from here – and even after a century of intensive field research, new species are continuing to be identified. That in itself is remarkable. But when you consider that it has about 4 per cent of the world's known dinosaur species, then you start to realise how very little we know about dinosaurs in most parts of the world.

Dinosaur Provincial Park is just one site representing a single complex of environments that covers less than 3 per cent of the time frame over which dinosaurs existed. Excluding their bird descendants, dinosaurs dominated the world for almost 150 million years, and it is doubtful that any single species was in existence for more than a few million years. They occupied every corner of the world and, like modern animals, different species would have been adapted to live in specific environments. Now if every environment and time period had as many dinosaur species as Dinosaur Provincial Park, then it is evident that there must have been tens of thousands of species that are not known.

In fact, one of the reasons so many new dinosaurs are being discovered these days is that scientists can identify the relative ages and ancient environments of rocks in different parts of the world. If our present knowledge of the dinosaurs from that time period, geographic area and/or environment is poor, then we can maximise our chances of finding a new species by going there to hunt for their remains. And this is just what people have been doing in recent years, in locations from China and Australia to Madagascar and Alaska.

There is of course a lot more involved. For example, not every rock type or site has the same potential for preserving fossils, and many palaeoenvironments would have been inappropriate for fossilising anything. Nevertheless, the example of Dinosaur Provincial Park shows that the potential for new discoveries is immense for other parts of the world that have so far been little explored for fossils.

Given the fact that we have likely discovered such a small number of the dinosaur species that must have lived in the past, the odds are against these few being the most unusual examples. New discoveries are constantly challenging our beliefs that previously known dinosaurs are weird. *Allosaurus* and *Tyrannosaurus* look pretty normal compared to crested or horned theropods, such as *Cryolophosaurus* or *Majungasaurus* – or even stranger new discoveries such as *Deinocheirus* or the various small feathered, flying dromaeosaurs of China.

The 'weirdness' of any animal is of course a function of our lack of familiarity with it, and books like this are critical for understanding that dinosaurs have adaptations, appearances and behaviours every bit as diverse as those seen across the full range of modern animals.

Foreword

Dinosaurs have long fascinated scientists and the general public alike, and there seems to be an unquenchable thirst for new information. *Weird Dinosaurs* is a book for the thirsty – packed with new knowledge and strange stories about recent finds that few readers will have heard before.

Philip Currie, MSc, PhD, FRSC
Professor and Canada Research Chair, Dinosaur Palaeobiology
University of Alberta

WEIRD DINOSAURS

INTRODUCTION: A NEW GOLDEN AGE FOR DINOSAUR SCIENCE

On a recent afternoon spent in London's Natural History Museum (NHM), I noticed that a life-size animatronic *Velociraptor* had finally sprouted feathers. They might have been a touch sparse compared to the plumage of counterparts in museums elsewhere, but it was encouraging to see them there. A few years earlier, the *Tyrannosaurs: Meet the Family* exhibition at the Australian Museum in Sydney had featured not only wonderful feathered fossils from China, but also models of shaggy carnivores that could have been swiped from Jim Henson's workshop yet were still creepy enough to strike fear into the hearts of gobsmacked kids.

Similarly, a model of *Velociraptor* relative *Deinonychus* at the Vienna Museum of Natural History in Austria is now lavishly feathered with beautiful, russet-hued plumage, small wings and a fetching mohawk, while the *Dinosaurs Among Us* exhibit at the American Museum of Natural History (AMNH) in New York is brimming with magnificent reconstructions that make it very clear our stereotypical conceptions of dinosaurs are dead and the boundary between these creatures and birds has become entirely blurred. So much has changed about our knowledge of dinosaurs – and in the seeming blink of an eye, too.

In 2014, my previous book, *Flying Dinosaurs: How Fearsome Reptiles Became Birds*, was published. It was the culmination of more than a decade of reporting stories about dinosaurs, and brought my time as a science writer back full circle to my postgraduate degree at the NHM. Writing that book, and talking to dinosaur hunters and museum scientists around the world about their astonishing discoveries, was a fascinating journey for me, and reawakened my love of these prehistoric enigmas.

But as I spoke to people about the Chinese work on feathered fossils, I realised something special was happening worldwide. From Siberia to Morocco and Alaska to Madagascar, new dinosaur discoveries were being made faster than ever before. There are so many new species and such a stream of fresh information that it is difficult for palaeontologists just to keep up with the research in their specific field of interest or the latest discoveries from their home nations. I've been lucky to experience some of the new finds myself on fossil digs and visits to museums in China, Mongolia and outback Australia. And alongside all these discoveries are fascinating tales of the people behind the research – their anecdotes from fossil digs and the historical legends of adventure.

More dinosaurs are being discovered right now than ever before – and not just a few more: the rate of discovery has been increasing nearly exponentially. The figures are fairly staggering: more than 50 new species of ceratopsian dinosaur, related to *Triceratops*, have been named since 2002 – many in North America, where *Jurassic Park*–inspired kids of the 1990s have grown into an army of new professional palaeontological recruits.

Having more feet on the ground in places that have historically been productive fossil localities is one way to find new fossils, but opening up whole new horizons is another – either novel slices of prehistoric time in the rocks, or whole new geographic locations. And this has been happening in earnest in the last two decades, most notably in China (where more than 150 new dinosaurs have been discovered since the early 1990s) and South America (where more than 40 titanosaur sauropods have been described in the last decade).

But, as you will see in this book, new fossils are now coming from all over, like never before – from Siberia, Australia, Alaska, Venezuela, Uzbekistan, Madagascar, South Africa, Romania, Mongolia, Malawi and even Antarctica. In many of these regions, growing wealth and development are helping homegrown palaeontology blossom, while local fossil hunters who studied in the West are now building their own teams and capabilities.

An abundance of dinosaur riches

It is extraordinary to note that the vast majority of species and what we know about dinosaur biology has come in the last few decades. A rough crunch of the figures suggests that around 50 per cent of all known dinosaur genera (genera are groupings of related species, e.g. the genus *Homo* includes the species *Homo sapiens* and *Homo neanderthalensis*)[1] have been found in the last 10 years, and nearly 75 per cent since 1990. Back then there were just 285 established genera;[2] in 2006 it had risen to about 527;[3] and currently it's about 1100, depending on the number of valid species you accept.[4] And this is just the figure for genera, which underestimates the number of actual species.

Back in 1993, when *Jurassic Park* was released in cinemas, it was bang up to date and based on the most recent science – but so much has changed since then that the public perception has struggled to keep up. Dinosaurs are no longer the green or grey, dim-witted, lizard-like creatures we thought they were before the 1980s, nor the scaly, reptilian predators we remember best from *Jurassic Park*. Today we know they were fleet-footed and often feathery, with sharp intellects and also strange behaviours, physical attributes and adaptations.

If you are not familiar with these 'weird' dinosaurs, then this book will open your eyes to a whole new prehistoric world. They include creatures that have snatched the headlines in recent years, such as *Spinosaurus*, a swimming, sail-backed predator from Africa that was larger than *T. rex*; and *Deinocheirus*, a huge ornithomimid dinosaur with massively long arms, the hump of a camel and a duck-like bill. Both of these were creatures first found long ago by German and Polish explorers, but were lost in the mists of time and the dust of the desert until new specimens were found and old quarries rediscovered. This is part of a recent trend in palaeontology of sleuthing out old fossil sites based on historical maps, notes and photographs, and it sometimes results in thrilling detective stories.

Even though we have likely so far only scratched the surface of dinosaur diversity, what we are finding is allowing experts to ask questions of a kind they have never been able to ask before. Large collections of specimens and species mean that these scientists can fill gaps in databases and in their understanding of the anatomy and evolution of dinosaurs in all the major groups. They are starting to discover new patterns in how some dinosaurs evolved in one region and fanned out to colonise other parts of the world.

Elsewhere, researchers are realising that the southern landmasses that once formed the supercontinent of Gondwana (South America, Australia, Africa, Madagascar, India and Antarctica) had their own unique fauna of carnivores and herbivores during the Cretaceous period. These discoveries tell us more than just about the dinosaur themselves, they give us clues to the movement of the continents and the ancient climatic conditions on our planet.

From the tropics to the poles

Once there was an idea that dinosaurs were cold-blooded and only thrived in the swamps and wetlands of tropical climes. But the more we look, the more we realise dinosaurs were found in as many different kinds of habitat as are birds and mammals today. Studies of fossils in Alaska, Antarctica and Australia are showing us that polar dinosaurs, probably warm-blooded and feathery, were thriving 70–100 million years ago in great polar forests, of which there is no modern equivalent. These Arctic dwellers include the newly described pygmy tyrannosaur *Nanuqsaurus* and the hadrosaur *Ugrunaaluk*.

New discoveries, such as the horned dinosaur *Spinops*, continue to be made in museum collections when old fossil are reappraised in the light of new knowledge, or backlogs of plaster-encased fossils are prised open many decades after the fossils were collected. Another weird ceratopsian discovered of late was *Regaliceratops*, from Canada, which had a crown-like halo of pentagonal plates around its neck frill. This species is one of many ceratopsians that had outrageous headgear, with spikes and plates sticking out in all directions like abstract works of art. This extreme ornamentation is beyond anything seen in nature today.

Ceratopsians, like many dinosaurs, also had unusual life histories. Unlike birds and mammals today, they may have started to reproduce before they had reached full adult size and appearance. Living animals may not be a good benchmark for judging dinosaurs, especially the biggest sauropods, which had to grow from babies the size of chihuahuas to adults bigger than 10 elephants in a mere handful of years.

Some of the dinosaurs you will meet in this book were undoubtedly very peculiar animals. But what do I mean when I say weird? Really, I mean dinosaurs that fall outside existing stereotypes, but dinosaurs are also weird in the sense that they display some traits that are utterly unfamiliar to us in living animals today.

Obviously, the enormous size dinosaurs reached – up to 70 tonnes and 38 metres long in a newly discovered species from Patagonia – is one of the very defining traits of the group. Dinosaurs were the largest land animals ever, reaching perhaps 10 times the size of the biggest land mammals (though not the blue whale, which may weigh as much as 180 tonnes).[5] As more sauropod discoveries come to light in many regions, from Egypt to Australia, experts are better able to answer questions about how they grew so huge, and why an evolutionary trend for gigantism existed in the first place.

While the herbivorous sauropods were rapidly growing ever larger during the Late Jurassic and the Early Cretaceous, feathered theropod carnivores were quickly evolving in the opposite direction. These relatives of *Velociraptor* were sprouting complex flight feathers and sequentially shrinking in size through evolution as they approached the appearance of the first bird. We can now piece together a remarkable evolutionary sequence through the 50 or so species of feathered dinosaur found as fossils. We now know that a number of very weird experiments in flight were going on at this time, too, as revealed by species such as the four-winged gliders *Changyuraptor* and *Microraptor*, and one of the weirdest dinosaurs ever discovered, *Yi qi* ('ee-chee'). This species, revealed in 2015, had both feathers and leathery gliding membranes like those of a bat. Even the experts who discovered it could barely believe this species had existed.

As entire unknown and unpredicted groups of dinosaurs have been discovered, some palaeontologists have come to believe that previous estimates of the true diversity of dinosaurs have been radical underestimates. Both Phil Currie in Canada and Xu Xing in China now believe there were tens of thousands of species of dinosaur, found in every imaginable habitat on earth, much as birds and mammals are today.

Birds, for example, which are the living descendants of the carnivorous theropod dinosaurs, number as many as 10 000 modern species alone. Across the Mesozoic, the non-avian dinosaurs persisted for about 180 million years, and it is clear that vast numbers of species must have come and gone over this time. In reality, even tens of thousands of total dinosaur species may be an underestimate. The fact that we know of fewer than 2000 species from fossils means we must be aware of just a small fraction of that diversity, which is biased towards the animals more likely to appear in the fossil record.

That record is patchy, as fossilisation is a fickle and implausible process. There are many parts of the planet where fossils would very rarely be preserved, such as uplands and areas without rivers, mountains, deserts and rainforests. We are missing many of the dinosaurs from these habitats, and whenever palaeontologists can discover new fossil beds that preserve these kinds of deposits, they open up a whole new window on a different environment. This happened in north-eastern China in the 1990s with feathered dinosaurs, and similarly exciting localities will continue to be found in the coming decades.

The incredible truth about the past

Not only are we finding new dinosaurs everywhere we look, but palaeontologists now have much more advanced technologies. They are using electron microscopes to study tiny features on the surface of fossils and reveal the colour of feathers; they are applying new techniques to find traces of collagen and other organic molecules, and even remnants of tissue and blood vessels; they are studying the chemical signatures in dinosaur tooth enamel to reveal the food they ate; and they are now even scanning fossils with a technique called laser fluorescence to reveal hidden details such as skin, feathers, scales and other soft tissues. Other digital analyses are allowing tests of the biomechanics of how sauropods walked and the aerodynamics of how other dinosaurs flew.

The science of palaeontology has advanced at a rapid pace in the past two decades, but Hollywood has been slow to catch up. Many would be surprised to learn that our current understanding of dinosaurs is nothing like the creatures depicted in 2015's *Jurassic World*. The most exciting new scientific developments include the fact that carnivorous dinosaurs were covered in feathers, and that we even know the colour of the plumage of some species. The new species are weirder than anything movie producers have been able to devise. We now know, for example, that not only would *Velociraptor* have had plumage, but it would have been decked out with fans of large feathers on its forearms, something akin to wings.

We have found these large wings on the fossil of a closely related Chinese dromaeosaur, *Zhenyuanlong*, revealed in 2015. While no fossils of *T. rex* – the most iconic dinosaur of all – have been found with feathers, we know that many related tyrannosaurids, such as the 9-metre-long Chinese species *Yutyrannus*, were covered in downy feathers, which suggests that *T. rex* itself had feathers.

Many moviegoers get their knowledge of dinosaurs from films and television. Not to show dinosaurs as we now know them to have appeared is spreading misinformation. Thankfully, feathered dinosaur toys and children's books featuring some of the unusual new species are beginning to appear and proliferate.

Perhaps the greatest crime, though, is denying people an incredible truth about the world around them – the fact that dinosaurs didn't become extinct when an asteroid hit the planet 66 million years ago. Sadly, the supersized and wonderfully alien creatures of the Jurassic and Cretaceous are now gone, but an echo of them lives on today in the thousands of modern species we know as birds. Chances are you are pretty close to one of the 400 billion avian dinosaurs that are flitting around, soaring on thermals, feeding or roosting right across the world right now.

Join me now on a voyage around the planet and into the distant past, where we will meet a series of genuine prehistoric oddities so curious that you'd be hard pushed to imagine them in your wildest dreams …

1

MONSTER FROM THE CRETACEOUS LAGOON

THE SAHARA, EGYPT

In 1911, a Bavarian palaeontologist found a partial skeleton in the Sahara Desert of a very peculiar carnivore, even bigger than T. rex, and with a huge sail on its back – but a wartime bombing raid blew the fossils to pieces, and the creature seemed forever lost until an unlikely chance encounter in Morocco in 2013.

Crabs scuttle along a tidal flat near the entrance to a great braided river system that is flowing out to sea on the north coast of Africa, 95 million years ago. Apart from the insistent buzz of insects, all seems quiet along the banks, as animals rest in the shade to escape the heat of the midday sun. A crocodile the length of a bus is sleeping, its tail poking out from vegetation. But things are not as tranquil below the surface of the wide, slow-moving river channel. A fizz of bubbles hints at the presence of giant, car-sized coelacanths and lungfish lazily moving through the waters. Nearby, a more energetic giant sawfish is waggling its several-metre-long rostrum, lined with barbed teeth, through bottom sediments as it hunts for prey.

None of these fish has noticed what is stealthily gliding towards them below the water with a flick of its great tail. Made obvious above the water by a great red sail, which slices through the surface, this killer moves almost silently and invisibly through the murky waters. It is an enormous carnivorous theropod, but it has very unusual adaptations for a dinosaur – including flattened, possibly webbed feet, a barrel-shaped body, and pits along its snout where pressure sensors detect movement in the water. The only semi-aquatic dinosaur we know of, it mostly preyed upon the giant fish that lived alongside it in these Gondwanan river systems, and rarely on other dinosaurs

Our first hints of this enigmatic creature came from Ernst Freiherr Stromer von Reichenbach, a Bavarian aristocrat who undertook a series of fossil-hunting expeditions into the Egyptian Sahara between 1901 and 1911. His fossil collector, Richard Markgraf, found tantalising clues to a massive carnivore – a creature even larger than *T. rex* – which Stromer named *Spinosaurus aegyptiacus*. It had features he was at a loss to explain, such as a 2-metre-tall sail and a long, almost crocodilian jaw packed with conical teeth.

Stromer studied the fossil back at his Munich museum, but many of the pieces were missing, leaving gaps in his knowledge. All hopes of learning more were dashed during World War II, when an Allied bombing raid in 1944 destroyed the museum and most of Stromer's Egyptian fossil trove, including all of the bones of *Spinosaurus*. All that remained were publications, field notes and a few sepia photographs. The book seemed to have closed on *Spinosaurus*.

But then in 2013, a young Moroccan–German palaeontologist, based at the University of Chicago in the United States, made a fortuitous discovery thousands of kilometres west across the Sahara from Stromer's original finds. His find was so strange and so compelling it made the cover of *National Geographic* magazine.

A series of lucky breaks

Nizar Ibrahim is sitting outside a cafe in the eastern Moroccan city of Erfoud in March 2013. There's a buzz in the evening air as locals head out to shop at the souk or gossip over mint tea, and tourists come back from the Erg Chebbi Dunes. Part of the reason for Ibrahim's trip out here this time, to the Moroccan Sahara near the border with Algeria, is that he's hoping to spot a fossil hunter he'd met here previously – a man who may be the key to uncovering important details about a very significant specimen. But Ibrahim, then aged 31, knows it's a long shot. He ran into the guy once before, in 2008, and aside from a somewhat fuzzy recollection of his face, all he can remember is that he has a moustache.

The story really began in April 2008, when a young Ibrahim was first in this oasis town after several weeks prospecting for fossils amid the stark beauty of the desert. 'We were collecting some really incredible fossils', he tells me. 'Thousands of specimens – remains of crocodile-like hunters and flying pterosaurs, and also dinosaur teeth and bone fragments. We found bits and pieces of *Spinosaurus*, but really nothing that would dramatically improve our understanding.'

He was resting in town after the dig when a fossil hunter brought him a cardboard box full of pieces that were clearly from a large animal but covered in sediment and thus difficult to make out.

But one unusual piece caught Ibrahim's eye – a section of dense, flat, blade-shaped bone. He thought it could be part of a massive rib – and, in passing, even wondered if this is what a broken *Spinosaurus* sail spine might look like look. Turning it around in his hands, Ibrahim noticed a red streak running down the middle of the cross-section.

Then in the midst of his PhD research, he decided to leave the mystery bones with his colleague, Professor Samir Zouhri at the Université Hassan II in Casablanca, but hoped he would one day return to study them. 'It was always on my mind, and I often thought it would be nice if I could identify these bones.'

In 2009, Ibrahim found himself at the Civic Museum of Natural History in Milan, Italy. His colleagues there, Drs Cristiano Dal Sasso and Simone Maganuco, showed him the partial skeleton of an odd dinosaur in the basement, which had been donated by an Italian fossil dealer in North Africa. They wanted to return it to its country of origin, but couldn't be sure it was Moroccan.

'They showed me the bones and my jaw just dropped, because I saw that this was a part of carnivorous dinosaur, which in itself is rare ... But also it had leg bones, strange claws and an almost complete foot. The claws were flat, which is very unusual in predatory dinosaurs, and the feet looked almost flat and paddle-like', he says. 'And there were ribs, backbones, tailbones, hipbones, pieces of the skull.'

Excited, his pulse racing, Ibrahim tried to piece the information together. But when his eyes rested upon what were obviously several long sail spines broken into pieces, he knew what this had to be. Then he looked at the spine in cross-section and saw a red line. 'I thought, wow, this looks familiar', he says. Not only did Ibrahim have a good idea this was *Spinosaurus* from Morocco, but the size, shape, colour, texture and sediment around the bone – as well as the telling red streak – hinted this was the same individual animal the moustachioed fossil hunter had brought him pieces of several years earlier in Erfoud. His mind whirring, Ibrahim pondered whether the guy had kept digging the same site, turning up more bones, which had found their way to Milan.

This would be a miraculous coincidence – it would mean he could confirm the first skeleton of *Spinosaurus* from Morocco, but also potentially return to the same spot to look for more of the individual, and then collect the all-important contextual information that would tell him the age and all the details of the environment the species had lived in. 'But to do that I had to travel back to Morocco and relocate the man', Ibrahim says. 'Now that's where our problems started.'

He hadn't entirely told the truth to the Moroccan colleagues he was travelling with in March 2013, including Samir Zouhri, as he knew it sounded ridiculous. Eventually he admitted his plan, but was embarrassed to confess that all he could remember was that the man had a moustache. 'Many men of a certain age in this part of Morocco have a moustache', he says. 'It really didn't narrow things down, and – to paraphrase what my colleague said, because he was not very impressed – that was a very poor starting point for our search.'

Nevertheless, they went to the border region and stopped at spots where they knew people were digging precarious tunnels hundreds of metres into the rock to find fossils. They chatted with locals, but couldn't find anything helpful, and were fast running out of time. Ibrahim had all but given up hope, and on the last evening was sitting in a pretty dejected state outside the cafe in Erfoud, sipping mint tea with Zouhri and also Dr Dave Martill, a palaeontologist from the University of Portsmouth in the United Kingdom.

'I was at my lowest point', Ibrahim says, 'and then I just see this person walking past our table and there was just something – like a déjà vu moment – and all my senses were on high alert. And the man had a moustache, of course, and I just thought, "This is the guy". It was so odd, I thought I was hallucinating'.

Never one to let an opportunity pass him by, Ibrahim stood up and darted after the man. A brief conversation in Arabic revealed it was the mystery man. He remembered Ibrahim and said that he had indeed found more bones at the site and sold them to an Italian dealer. 'Everything suddenly fell into place. It was just an incredible moment.'

The man remembered for what he'd lost

A little over a century earlier, in January 1911, German fossil hunter Ernst Stromer had also struck gold in North Africa – but this time in Egypt, almost on the opposite side of the continent. Though Egypt was then part of the British Empire, Germans had played an important role in the European exploration and mapping of its Western Desert. Stromer was following in that tradition when he had first visited Egypt in 1901, to explore the El Fayoum region, south-west of Cairo, which is rich with mammal fossils from the Eocene–Oligocene boundary, about 34 million years ago. Egypt was a fertile ground for finding fossils – Baron Nopcsa of Transylvania (see chapter 3) had visited the Fayoum Oasis in the years before World War I, and corresponded with Stromer on a number of occasions.

On his second visit, Stromer arrived in Alexandria in late 1910, and from there planned an expedition via camel train to the Bahariya Oasis, 290 kilometres south-west of Cairo. Stromer made many of his discoveries with the help of his Egypt-based fossil collector and friend Markgraf. Although Markgraf missed this particular expedition through illness, he returned to Bahariya in the years that followed and collected many of Stromer's important finds.[1] Stromer had a specific objective – he was setting out to find fossils that would back up his notion that mammals originated in Africa – and we know a great deal about his expedition because his journals still exist. 'He recorded his activities in fastidious – one might say numbing – detail', says author William Nothdurft, who gives an English account of Stromer's adventure in his book

The Lost Dinosaurs of Egypt. 'Each journal entry is dated, his location given in precise coordinates, the times noted to the minute.'

On 18 January 1911, after many days of largely fruitless searching in the area between the Fayoum and Bahariya oases, Stromer struck gold at the north-western corner of Bahariya at a spot called Gebel el Dist. But it wasn't at all what he was looking for: he'd found a dinosaur graveyard, and in just a few hours turned up a thighbone, vertebrae, part of a rib and 'a gigantic claw'. With what Nothdurft describes as characteristic understatement, Stromer scribbled: 'Apparently these are the first of Egypt's dinosaurs and I have before me the layer that contains land animals'.

But he had no idea how to 'conserve such gigantic pieces ... or transport them to Fayoum'. Having come to find small mammals, Stromer lacked the tools or means to excavate them and transport them back to Munich via Cairo. Eventually, with the help of Markgraf, the array of dinosaur fossils was packed into eight crates and shipped to Munich – and Stromer arranged for Markgraf to return to Bahariya, to continue the prospecting and excavations.

After the onset of World War I in 1914, it became impossible for Stromer to ship many of his remaining fossils to Munich. With the help of Swiss palaeontologist Bernhard Peyer, he finally managed to get hold of them in 1922, but they arrived damaged and broken into hundreds of pieces.

Between 1911 and 1914, Stromer and Markgraf had found the remains of three giant carnivorous dinosaurs – *Bahariasaurus*, *Carcharodontosaurus*[2] and *Spinosaurus* – and one herbivorous sauropod, which Stromer called *Aegyptosaurus*. They also found small pieces of another sauropod and some carnivores that were more difficult to identify, as well as snakes, crocodiles, turtles and marine reptiles. *Spinosaurus* itself was described in a paper in 1915, but Stromer continued to describe the material until 1936 – it became his life's work.

Like *Spinosaurus*, *Carcharodontosaurus* was also probably bigger than *T. rex*. And following a 2007 study from researchers at the

Royal Tyrrell Museum of Palaeontology in Canada (inventively titled 'My theropod is bigger than yours ... or not: estimating body size from skull length in theropods'),[3] it held the title of largest carnivorous dinosaur of all – until the discovery of the new, much more complete *Spinosaurus* specimen in 2013. According to the estimates in the 2007 study, published by François Therrien and Donald Henderson, *Carcharodontosaurus* would have reached 13.26 metres long and around 15.1 tonnes, while *T. rex* only stretched to 11.97 metres and 9.11 tonnes.

Carcharodontosaurus was a predator that looked similar to *Allosaurus* or *T. rex*, but was possibly heftier and meaner. Perhaps *T. rex*'s enduring place as the 'tyrant king' of the dinosaurs in most people's minds has to with the fact it is a prominent species from the United States – and that more than 50 specimens have been found, many largely complete. Though Stromer's *Carcharodontosaurus* specimen was one of the ones later destroyed, the University of Chicago's Paul Sereno found another skull in Morocco's Kem Kem fossil beds in 1995 – and then later a further specimen in Niger, which was described as a new species, *C. iguidensis*, rather than the original *C. saharicus*.

But the sum total of remains is still paltry compared to the impressive selection of *T. rex* skeletons on display around the world. It is interesting to note that Henry Fairfield Osborn's description of *T. rex* in 1905 achieved massive fame globally, but Stromer's descriptions of *Spinosaurus* in 1915 and *Carcharodontosaurus* in 1931 sank almost without trace. In 1915 at least, with the advent of World War I, people had other things on their minds.

The fact that two of the largest carnivorous dinosaurs ever known lived near each other in the same region is interesting, too – could they have both grown so large because they sometimes faced off against each other in battles for territory around the water's edge? Or was there another factor at play here in this North African region of Gondwana? At this stage nobody knows for sure.

Bahariasaurus was another allosaur relative, in a similar size range to *T. rex*. Unfortunately, we know very little about *Bahariasaurus* or the sauropod *Aegyptosaurus*, as no more significant specimens have been found since they were also destroyed in 1944.

The confusing number of large carnivores at Bahariya came to be known among experts as 'Stromer's three-theropod riddle'. In a typical dinosaur ecosystem of any given time period, one large carnivore is the apex predator – for example, in the Late Cretaceous there was *Tarbosaurus* in Mongolia, *Giganotosaurus* in Argentina and *Tyrannosaurus* in North America. Different huge carnivores come and go, but rarely do they overlap in the same place at the same time. At Bahariya, however, Stromer had seemingly found three apex predators all in the same environment. How was that possible, and what could they possibly have been eating?

Many of the Bahariya remains are of marine animals, and *Aegyptosaurus* was not an especially large sauropod, despite being related to the titanosaurs of Patagonia. Three – or more – huge carnivores required a lot of meat, and the ecosystem that provided this would have needed to be a bountiful one, but the fossil record provided little clue as to where this food was coming from. The answer to that question would come, but not until long after Stromer's life had ended following a series of crushing setbacks, including the loss of all his fossils and two of his sons. After Stromer's Egyptian fossil trove was destroyed in Allied bombing raids in 1944, the Gebel el Dist dig site remained all but forgotten. Even Stromer himself never returned there after 1911.

The tidal giant at the Bahariya Oasis

The first part of the answer to the three-theropod riddle would come almost 90 years after Stromer's first finds, when a new team of explorers set out to the Bahariya Oasis, deep in the Sahara Desert. In January–February 2000, this US team, funded by a TV documentary crew, headed out on a seven-week expedition, hoping to rediscover Stromer's original sites.

As the expedition's chief palaeontologist, Dr Josh Smith, told Nothdurft: 'There were several reasons to consider Bahariya ... The first was no paleontologist had been there since Stromer. Second, very little is known about the terrestrial animals of Late Cretaceous Africa, beyond what Stromer and Markgraf found in Egypt, so there was an opportunity to make a real contribution to the science. Third was that everything Stromer found had been lost'. This loss seemed a gross injustice, said Smith. 'We decided we wanted to try to recapture and rebuild his legacy.' Ken Lacovara, now at Rowan University in New Jersey, was chief geologist on the expedition.

The discovery of the site had been serendipitous – Smith, then a 29-year-old graduate student at the University of Pennsylvania, had mistakenly entered coordinates into his GPS device on a preliminary survey the year before. That error had led him to chance upon an unusual bone poking out of the ground, and they now returned to that site to investigate.

The conditions in the Sahara can be very challenging – Lacovara and Smith's team had to contend with great sandstorms that lasted for days and the constant threat of dehydration. They also ran the risk of running into venomous camel spiders, centipedes, death stalker scorpions (the world's most venomous) and pit vipers. But there was time for a little fun, too. Lacovara – who plays the drums and had studied music at university as well as geology – says that on several occasions he got to play spontaneously alongside travelling Bedouin musicians in a village near the dig site.

Though they had taken a gamble on the expedition, they were not disappointed – they retrieved nearly 7 tonnes worth of fossils, including the 1.7-metre-long upper arm bone of a new and massive sauropod dinosaur – *Paralititan stromeri*, a titanosaur that would have weighed perhaps 60 tonnes and been 27 metres long.[4] Other parts of the partial skeleton included huge ribs, vertebrae, shoulder bones, bits of forelimbs and even what may have been armour plates from the skin. As with Late Cretaceous sites in Morocco and Tunisia, the remains of coelacanths and other fish, turtles and crocodiles were also recovered.

The sandstone and organic mudstone rocks in which the fossils were found had ripples and other evidence that waves had once gently lapped against them. While today the Bahariya Oasis is in the middle of one of the world's greatest deserts, 95 million years ago it was tropical coastal site, with mangrove swamps and vegetated tidal channels, something like those found today in the Florida Everglades. The genus name, *Paralititan*, means 'tidal giant', while the species name honours Ernst Stromer.

'The discovery of a huge sauropod, especially in a near-shore environment, is of great interest', palaeontologist Hans-Dieter Sues, who was not involved in the expedition, told reporters when the finds were announced in 2001. 'The Egyptian material represents a fauna that is widely found across North Africa, all the way to Morocco in the west, but it documents the dinosaurs much better than the other occurrences, which have usually only yielded isolated bones and teeth.'

The find excited other palaeontologists, as the rediscovery of Stromer's original Bahariya Oasis site offered a new window onto the dinosaurs of North Africa – and the work was then unusual for having pieced together the ecosystem and wider environment through detailed study of the sedimentary layers, trace fossils and remains of plants and other animals, including invertebrates. It was this work, led by Lacovara, that provided the evidence Bahariya had been a lush and productive coastal site in the Late Cretaceous. This was a period of global 'hothouse' climate, when there was little difference in temperature between the poles and the equator, and sea levels were high.

In 2001, and again in the following years, Lacovara's team returned to collect more of the other plant and trace fossils, which he describes as less glamorous but essential to scientists who want to place dinosaurs in their ecological context and go beyond mere trophy hunting.

The discovery that there had been massive herbivorous sauropod dinosaurs at Bahariya – which would have provided food for the terrestrial carnivores *Bahariasaurus* and *Carcharodontosaurus* – went part of the way to answering the three-theropod riddle, as did the fact that these environments had been lush coastal river systems. But it was the fact that *Spinosaurus* hunted and lived within the rivers themselves, as finally confirmed in 2013, that would really explain how three large carnivorous dinosaurs could have coexisted alongside one another.

A life spent in the waterways

Martill describes Ibrahim's ridiculous good luck with the Moroccan fossil hunter as an absolute case of serendipity. A long drive and a steep climb the next day, and Ibrahim, Martill and Zouhri were led to the site from which the *Spinosaurus* had been extracted. They began to poke around in the dirt that had been dug out of the hole and uncovered more bits and pieces. 'We uncovered some of the spine and I found quite a few of the teeth. He'd accidentally thrown them away', Martill says.

Morocco's Kem Kem fossil beds are fairly soft and relatively easy to dig, Martill adds, but – as in Egypt – the logistics are very challenging. It's unbearably hot later in the day. You are in remote locations and have to make sure that once you've gone 30–40 kilometres off the sealed roads, you have provisions in case of a breakdown. Depending on the political situation, parts of North Africa are unsafe places for Westerners to work – particularly Mauritania and Mali, but also Tunisia, Libya and Egypt in recent years.

Nevertheless, it's worth the effort, as the fossils can be abundant. The Kem Kem beds are layers that were mostly deposited in a very big, braided and meandering river system, comparable to the Congo River today, with lots of sandbars and other features. The remains of this river system have been traced across North Africa from Tunisia and Algeria, down to Mauritania, more than

1000 kilometres south. In these deposits there is evidence of flood events where the banks have collapsed into the river, and it's in these that palaeontologists find numerous bones of fish, and occasionally bits of terrestrial animals such as *Carcharodontosaurus* and *Spinosaurus*.

Several months later, in 2013, Ibrahim ended up with all the bones from Milan and Casablanca – and the new scraps from Morocco – in Chicago, where he could piece them together, with help from his co-worker, veteran dinosaur hunter Professor Paul Sereno. And, for the first time ever, they were able to have a good stab at reconstructing a whole *Spinosaurus*. To do this, they also used photographs of Stromer's original specimen and a series of other isolated fossil bones found over the years, to help fill gaps in the skeleton (a move that was derided by some critics as unscientific, but was nevertheless instructive).

What Ibrahim, Martill, Zouhri, Sereno and their Italian colleagues have been able to piece together is a completely novel kind of dinosaur, which they revealed in a paper published in the journal *Science* in late 2014. And it's an animal that looks very different from how Stromer imagined it, based on his partial skeleton from 1911. *Spinosaurus* was certainly huge, and at 7–9 tonnes and 15 metres long, bigger even than *T. rex* or *Giganotosaurus* (both about 12 metres). But unlike the *Spinosaurus* depicted in 2001's *Jurassic Park III*, it would not have walked upright on long hind legs and may not have taken on predators of *T. rex*'s ilk.

But its size is not the only interesting thing about it. All the evidence points to *Spinosaurus* having been semi-aquatic; apart from the living bird members of the theropod lineage, such as penguins, it is one of the only swimming dinosaurs yet discovered. Other, earlier, spinosaurs, such as *Baryonyx* from the United Kingdom, *Irritator* from Brazil and *Suchomimus* from Niger, were fish eaters that hunted in riverine and coastal environments, but none appear to have been animals that spent most of their time in the water.

'It's the first dinosaur that shows clear anatomical adaptation for a life largely spent in the water', Ibrahim says. 'The skeleton basically has "water loving" written all over it. You have these long slender jaws like a crocodile for rapid closing underwater to catch slippery prey; and conical teeth, which are very different from the typical predatory dinosaur teeth, which are knife-like and serrated for robust bone crunching. These jaws are perfectly suited to catch large, slippery prey – big fish. The nostrils are far back on the snout, not at the tip like in other predatory dinosaurs. This would have allowed the animal to plunge the jaws underwater, maybe while wading, and still breathe, because the nose opening is above the water. The tip of the snout has all these big openings, which we think are more or less analogous to what we see in alligators and some other crocodiles, where these openings house pressure receptors, so in murky or dark water it can still detect movement.'

And that's just its head. *Spinosaurus* also had a long neck, like a heron or a stork, a barrel-shaped body, and small, powerful hind legs. It may even have walked some of the time on all fours, as its centre of gravity meant it stood much less upright than other theropods. Most theropods have low-density bones full of air pockets similar to birds, but *Spinosaurus*'s are much denser, resembling the bones of aquatic mammals such as hippos or prehistoric early whales.

'It would not have been incredibly agile on land', Ibrahim says. 'We also have these very strange claws; and when you look at the foot, it's not far-fetched to imagine that this was maybe a webbed foot. It has a fairly flexible tail, which is also quite unusual ... This is an animal that has all the tools and all the equipment to go after large aquatic prey.'

Its huge size perhaps reflects the fact it shared its rivers with huge 12-metre-long crocodiles. The fossils, as yet, show no evidence of feathers, so it's possible that *Spinosaurus* was scaly rather than fluffy. Other large carnivores with feathers may have used them for display or insulation – neither of which *Spinosaurus* would have needed them for. In fact, scientists believe it was a territorial animal, and the giant sail, supported by the large neural spines, was a display feature.

'Imagine these big river systems with different individuals fishing', Ibrahim says. 'Many animals today are quite territorial; they'll take control of a certain area or branch of the river and will defend it. For animals that live in the water or spend a lot of time in the water, it's difficult to convey this information and communicate things like their size, because they're mostly submerged. So the sail would obviously be visible, because that would be the only thing sticking out of the water. It would be a great way of telling other animals how big it is.'

The discovery that *Spinosaurus* spent much of its time hunting in the rivers helps explain how the three carnivores of Bahariya were dividing and sharing up the environment between them. 'We're talking about this huge river system, and most of the biodiversity is concentrated inside the river. So there's certainly abundant food in the rivers. We're talking car-sized coelacanths and giant sawfish and really, really big lungfish, I mean true giants', Ibrahim says. 'It explains how *Spinosaurus* could have existed in this environment with other large predatory dinosaurs.' They were not in direct competition with one another, as they had all evolved to be highly specialised in different parts of the environment.

It took 95 million years to uncover the first *Spinosaurus* fossil and another century for the second skeleton to be found, but with what researchers – such as Ibrahim and Smith – have learnt in the past 15 years about where best to find Saharan dinosaurs, we can hope we don't have to wait so long for the next exciting discovery.

A legacy restored

As part of his research into *Spinosaurus*, Nizar Ibrahim – who had himself grown up in Berlin – travelled home to Germany to look at Stromer's archives. He went to the family castle in Bavaria and met with Stromer's granddaughter, Rotraut von Stromer-Baumbauer. 'She'd heard a lot from her father and had all these old records, and it was very moving to listen to these stories.'

The more Ibrahim learnt about Stromer, the more he realised he was a fascinating character he could identify with. Like many gentlemen scientists of his day, Stromer was very well read and was a prolific writer, on topics as diverse as coral reefs and politics in addition to palaeontology.

After the Nazis rose to power in 1933, the political climate began to deteriorate rapidly for Stromer, then a professor of palaeontology at the Bavarian State Collection of Palaeontology and Historical Geology, housed in a museum at Munich's Alte Akademie. Not only did he refuse to join the party, but he kept close ties with Jewish friends and, on occasion, even spoke out against the regime. In 1937, aged 65, he was forced to retire, but continued some research as a fellow of the Bavarian Academy of Science.

In 1940, Karl Beurlen, a respected younger palaeontologist and ardent Nazi became the director of the collection. As the war progressed and the museum came under increasing threat of air raids, Stromer repeatedly demanded that Beurlen have its treasures moved, but Beurlen refused, Ibrahim says. 'Most major museums in Germany were moving their scientific and art collections to places outside the big cities, in castles or underground mines, because devastating Allied bombing campaigns had destroyed large parts of many cities.'

Very early in the morning on 25 April 1944, Stromer's worst fears were realised. Two hundred RAF Lancaster bombers littered the city with bombs, destroying the central railway station and setting 7000 buildings ablaze. Within a few hours, the Alte Akademie housing Stromer's incredible Egyptian fossil trove – including the only specimen of *Spinosaurus*, mounted in one of the galleries – had been reduced to smouldering ruins.

'He discovered an entire lost world – it's really not just a couple of dinosaurs', Ibrahim says. 'He'd also found the best skeleton of the giant predator *Carcharodontosaurus*, and nothing even close has been found since. Just a few bits and pieces of skulls and teeth, but nothing like Stromer's fossils. You can just imagine what it must have felt like to him losing all these things in a single night. He talked about it until the end of his life. It was a very traumatic experience for him.'

Even more traumatic, of course, was the loss of his sons. Stromer had spoken his mind fairly openly in the preceding decade, and that was a dangerous thing in Nazi Germany. His standing as an aristocrat meant he avoided direct harassment from the Gestapo, but all three of his sons were drafted into the army and sent on dangerous postings. His oldest, Ulman, was killed in 1942; and his youngest, Gerhart – who was just 18 – was killed in southern Germany two weeks before the end of hostilities in 1945. Wolfgang was feared dead, but after the war Stromer found that he was in a Soviet prison camp. He returned home in 1950, two years before Stromer himself passed away.

Ibrahim felt a connection with Stromer, who was a more likeable figure than many of his contemporaries. 'He was a good person and graceful in his work and not someone looking for undeserved attention … Taking all this together gave me great insight. The whole *Spinosaurus* thing became something much bigger. It was no longer just trying to track down the fossil in Morocco, it was also trying to understand more about the animal and also – last but not least – restoring Stromer's legacy. Here was this forgotten scientist, one of the true greats of palaeontology, but who was remembered for what he had lost rather than for what he had found.'

Ibrahim says he'd long dreamt of following in Stromer's footsteps and rediscovering his lost world of Saharan dinosaurs, with *Spinosaurus* the centrepiece. 'It's kind of the most bizarre and the largest predator. To most people, carnivorous dinosaurs all look pretty similar … but with *Spinosaurus* I knew that we had something really, really bizarre.'

The rediscovery of this incredible dinosaur made it to the

cover of *National Geographic* magazine in October 2014 – and an exhibition at the National Geographic Society, in Washington DC, featured a life-sized replica of the skeleton created from the 3D digital model the team had been able to reconstruct.

'I have thought about what it would be like if I could bring Stromer back from the dead', Ibrahim says, 'and tell him all these things, and say, "Look, your *Spinosaurus* is on the cover of *National Geographic*!".' He may have taken some satisfaction in knowing that his grandchildren and great-grandchildren saw his scientific legacy restored. Having resolved this great mystery about *Spinosaurus*, Ibrahim now has his sights set on getting a national museum with a research collection built in Morocco, and having capacity there for fossil-hunting expeditions led by local scientists.

While North Africa has long been a site of dinosaur discovery, 20 years ago it would have been impossible to predict the incredible fossil riches that have since poured from China. From exquisitely preserved feather impressions to implausible evolutionary sidesteps, the dinosaur finds have been nothing short of astounding.

2

ALL HAIL THE DINO-BAT

HEBEI PROVINCE, CHINA

When is a wing not a wing? Since 1996, nearly 50 new feathered species of dinosaur have come from China. Experts thought we had a good idea of the diversity of these fluffy and fuzzy carnivores, and how they used their feathers, but one recent find was so unexpected it left palaeontologists scratching their heads, and hinted at how little we really knew.

The Shandong Tianyu Museum of Nature in Pingyi, China, is the place to go if you want to experience China's dinosaurs in all their weird and wonderful glory. Five hundred kilometres south-east of Beijing, it's the largest museum of its kind in the world, with fossils of more than 1000 complete dinosaurs, 2300 early birds and plenty of creatures that bridge the boundary between the two groups. Arranged in 28 halls in three nondescript-looking buildings, these numerous spectacular specimens merely hint at the treasures behind closed doors, where a backlog of new feathered dinosaur finds is stacked up waiting to be studied and named.

Species described from the collection here include *Tianyuraptor*, a dromaeosaur; *Tianyulong*,[1] one of the first known feathered ornithischian dinosaurs;[2] and the four-winged flying dinosaurs *Anchiornis* and *Xiaotingia*.[3] Many museums have single specimens of dinosaur species in their collections; Tianyu has about 250 fossils of *Anchiornis*, hundreds of *Microraptor* and several hundred other dromaeosaur specimens. 'You don't need to do too much to dig up fossils in China', local palaeontologist Wang Xiaoli told a reporter from the *New York Times*. 'When the wind blows, they reveal themselves.'[4]

Not much surprises the scientists and technicians who work on the geological riches that pass through these doors, but one fossil sold to the museum in 2007 had them stumped. 'Over the last 20 years I have discovered many dinosaurs, so, to a degree, I am used to finding strange species', says Xu Xing, China's most celebrated palaeontologist, based at the Institute of Vertebrate Palaeontology and Palaeoanthropology (IVPP) in Beijing. 'But still when I saw this particular discovery I was shocked. It makes you believe there is still a bigger potential in the future that you can find even more bizarre dinosaurs. It tells you how incomplete the fossil record is.'

Many of the treasures that arrive at the museum have been dug up by amateur fossil hunters, farmers who supplement their meagre income by excavating the remains of dinosaurs and early birds that they sell on to dealers and institutions. It has theoretically been illegal to buy fossils collected by amateurs since 2008, but the scientists admit that the legality of how these specimens arrive in Chinese collections is somewhat of a grey area. Better that priceless fossils are available for science, they argue, than lost to the living rooms of wealthy private collectors.

The majority of feathered dinosaur fossils come from the north-eastern province of Liaoning, near the border with North Korea, but some – such as one specimen that had experts perplexed – come from the neighbouring province of Hebei. This is a relatively new locality for fossil hunters, dated to the Mid- to Late Jurassic, 160–165 million years ago. 'The area that can produce feathered dinosaur fossils is not only the western part of Liaoning Province but also the northern part of Hebei and south-eastern part of Inner Mongolia', says Xu. 'Fossils in northern Hebei are not as rich as Liaoning, but often they are well preserved. In some cases, feathers and soft tissues are even better preserved.'

The new specimen was found by a farmer, Wang Jianrong, who dug it from a fossil quarry near the village of Mutoudeng in Qinglong County. Here Hebei's small mountains, deep valleys and rolling foothills are pocked with pits dug by farmers searching for fossils. Wang may not have realised he'd found something important to start with, as by the time the specimen arrived at the museum it was broken into several pieces and parts were missing. 'He had cracked open the rocks', Xu says, 'but when he saw the bone, stopped removing more rock and just collected them. He glued them back together, but there are still some pieces missing'.

Despite being based in Beijing, Xu has an ongoing collaboration with the museum in Pingyi and goes several times a year to study new fossils and work on a series of projects. The world's most prolific living dinosaur hunter, he has been involved in the discovery and naming of approximately 60 dinosaur species, from the magpie-sized, four-winged flyer *Microraptor*, found in 2000, to the giant 9-metre-long fluffy tyrannosaur *Yutyrannus*, described in 2012.

When I speak to Xu over the telephone in his office at the IVPP, he says he has at least 10 species of dinosaur he is about to name in new journal publications, and his office is stacked with specimens waiting to be studied and described. 'There are two fossils behind my seat right now and to my right a few more', he says. 'It's difficult to find anywhere to put your feet.'

He first saw the strange new specimen in 2009, shortly after it arrived at the museum. It was broken and partially covered with rock, so he could tell it was a feathered dinosaur, but not which group it belonged to. He thought there was a good chance it was potentially interesting, but decided first to focus on low-hanging fruit in the museum's collection, the kind of fossils where 'you can tell immediately what they are and what the implication is'.

Xu and the museum's founder and director, goldmining magnate Zheng Xiaoting, were busy working together and studying other fossils. When they finished with these specimens, they had more time to go back and look at the specimens they had initially skipped over. Xu picked up the fossil and, running his hands over it, decided there was definitely something strange about it. In 2013 he sent his technician Ding Xiaoqing to the museum and asked her to prepare it by cleaning the overlying rock away.

'Strange wing' is revealed

Several months of work later and – despite the fact that little of the specimen below the ribcage was present – Xu could see from the distinctive skull shape and the hand that it was a scansoriopterygid ('climbing wings') dinosaur. This feathered group includes the species *Epidexipteryx* and *Epidendrosaurus* (previously called *Scansoriopteryx*), which have forward-slanting teeth and weird fingers. Discovered in 2002 in Liaoning, *Epidendrosaurus* was thought to have been a sparrow-sized climber and tree-dweller, with an elongated third finger, possibly for the purpose of skewering insects hidden inside tree hollows, much as the aye-aye of Madagascar does today. The slightly larger, pigeon-sized *Epidexipteryx*, found in Inner Mongolia in 2008, was weirder still. Like *Epidendrosaurus*, it was covered in downy fuzz, but it had four long, ribbon-like feathers emerging from its tail and its third finger may have been half as long again as its entire body.

One difficulty with the new specimen from Hebei, was that it wasn't like the majority of feathered dinosaur fossils from Liaoning, in which bones and feathers are clearly preserved and easy to discern: alongside the clear remains of bones in these flattened, road-kill-like remains are improbable details very rarely fossilised, such as exquisite fans of feathers, internal organs and gut contents. In contrast, in Hebei the fossils tend to be highly compressed and in some cases represented by mere imprints rather than bony-looking, three-dimensional structures. In this case, it was difficult for the experts to tell if the non-bony features represented something unrelated to the skeletal system, such as ligament or muscle.

As with other scansoriopterygids, the feathers of the new creature were short and brush-like, and would have covered its whole body; scansoriopterygids lacked the vaned, 'pennaceous' feathers that are found on modern birds and many theropod dinosaurs. More of these fuzzy feathers were preserved on another piece of the fossil surrounding part of a hind limb and foot, and an analysis with an electron microscope revealed the presence here of melanosomes – tiny packages of pigment that are responsible for feather colour in living birds.

Furthermore, as Ding slowly removed the matrix of rock encasing the fossil, the experts noticed two rod-like structures near the wrist that were like nothing they had ever seen before. 'This confused us for quite a long time', Xu says. 'It was apparently part of the skeletal system, but something you never see in other dinosaurs.'

At this time, Xu's colleague Corwin Sullivan, a Canadian scientist also based at the IVPP, was working on a book with a section on gliding and flying animals, and noticed a sentence about flying squirrels that led him to an image of one of these creatures from a Japanese photographer. This photo showed the animal's 'styliform elements' – rods of cartilage that protrude from the wrists of some gliding mammals. They allow them to hold out their patagia, membranes of skin that extend from their wrists to their ankles allowing them to fly. Something similar is also seen on the feet of bats and the arms of pterosaurs.

Sullivan says the first time he saw the fossil was in early 2014, when he, Xu and American palaeontologist Jingmai O'Connor were visiting the Tianyu museum. 'What really puzzled us were these structures extending from both wrists that looked very much like bones on the slab, but of course theropods aren't supposed to have bones in this position. I remember standing around with Xu, Jingmai and Zheng trying to work out whether these were ligaments or some other kind of soft tissue, but we couldn't come up with anything that really made sense.' (Later humorous suggestions as to the purpose of the rods included that they might have been used as ski poles or giant chopsticks.)

Following this, Sullivan returned to Beijing and in the course of his research came across the sentence in the book and the picture. 'When I read that thing about flying squirrels, the fossil just flashed into my mind', he says. 'The idea was hugely exciting ... but I reined myself in, thinking there were probably a dozen reasons this couldn't be the same structure.' As he began to research it in more detail and discussed it with the others, however, he realised he was on to something.

Modern birds do in fact have a small skin flap in front of the elbow called a propatagium, which forms part of the wing but is covered by the much larger flight feathers. 'You can imagine that if this propatagium was also present in theropod dinosaurs that are the ancestors of birds, then what's going on in the scansoriopterygids is that it has expanded to form a wing membrane', Xu says. 'Finally we noticed there were patches of membrane preserved near the styliform elements. We suddenly realised this was a really bizarre dinosaur with kind of pterosaur- or bat-like wings. We all got so excited, and we just couldn't believe that this had happened, particularly because I had devoted so much time to the study of feathered dinosaurs. Feathered wings in my mind always had beautiful, big flight feathers. These are the major components of dinosaur wings. And here we had something totally different. I couldn't believe it, but the fossil was real and was it front of my eyes and I could touch it, so I had to believe it was true.'

Working with palaeontologists including Sullivan, O'Connor, Zheng and Xing Lida, based at the China University of Geosciences, Xu began to amass the exceptional evidence they would need to back up their extraordinary claim. Fake fossils, or chimeras, have been a big problem in China since the late 1990s, particularly since a high-profile composite of a number of birds and dinosaurs dubbed *Archaeoraptor* was heralded as a 'missing link' in *National Geographic* in 1999. Xu's expert eye is rarely fooled by inauthentic specimens, but he realised that a bat-winged dinosaur was so improbable that other palaeontologists might have doubts if his team hadn't done their homework.

Before they published a paper in *Nature* in April 2015, they did a variety of analyses to prove, for example, that the fossilised styliform element had a chemical composition suggesting it had originally been made of bone or cartilage. Crucially, the fossil had been prepared out of the rock at the Shandong Museum, so the experts themselves knew it could not be a chimera. 'I am 100 per cent sure this is not a faked fossil', Xu says. 'Sometimes I am not so certain if a fossil is real, or I think something may be wrong with it, but in this case I am quite certain.'

Furthermore, because of their connections with the network of fossil dealers and farmers in this region of China, they were able to pin down Wang as the original discoverer and then visit the specific region within the quarry from which he'd extracted the specimen, confirm its provenance, and collect other data such as the geological layer and age.

The team called the species *Yi qi*,[5] ('ee-chee') which means 'strange wing' in Mandarin. Not only was it remarkable for being the first dinosaur with bat-like gliding membranes, but it also had the equal shortest scientific name of any species, an honour it shares with an actual bat, *Ia io*. (In case you're wondering, the longest name appears to be *Parastratiosphecomyia sphecomyioides*, an Asian soldier fly.)

The announcement of the species was greeted with a mixture of bemusement and palpable excitement by a palaeontological community that had become somewhat desensitised to the feathered dinosaur riches pouring from China.

In an accompanying commentary, also published in *Nature*, palaeontologist Kevin Padian wrote that, following the discovery of *Sinosauropteryx* in 1996, 'the picture of the evolution of feathers and flight has become richer and more complicated as other feathered dinosaurs have been discovered, seemingly on a monthly basis. But things have just gone from the strange to the bizarre'.[6] Daniel Ksepka a palaeontologist from the Bruce Museum in Connecticut, described the discovery as 'refreshingly weird', adding that 'palaeontologists will be thinking about *Yi qi* for a long time'.[7]

The reign of the feathered dinosaurs

In the early 1990s, nobody could have predicted the incredible flurry of dinosaur discovery that was about to begin in China. Something like 160 species have been described since then, and the rate at which these new discoveries are being made may not even have peaked. By that time, most experts had accepted that birds were not only the descendants of dinosaurs, but that they were actually small, feathered, flying theropods. But few thought we would ever find non-avian dinosaur fossils with feathers.

That's why palaeontologists were astounded when a fluffy, 1.5-metre-long dinosaur called *Sinosauropteryx* was found by a farmer in Liaoning in 1996. There are now around 50 species of dinosaur for which we have direct evidence of feathers. Some have halos of fluff or beautiful fans of flight feathers delicately traced into their remarkable fossils; others have a distinctive pygostyle tailbone, which would have been an attachment point for feathers; still others have bumps along their forearms – 'quill knobs', which are where feathers attach to ligaments in the wings of modern birds. Most of these species come from Liaoning, Inner Mongolia and Hebei in China, but a handful are from Mongolia proper, Burma, Madagascar, Germany, Canada and Russia.

Together, these fossils show us that the majority of theropod dinosaurs all over the world would have had feathers. More than 90 per cent of fossil sites globally preserve only bones. Regions such as Liaoning, which preserve a Pompeii-like world of dinosaurs in fine-grained sediments of volcanic ash, many of the fossils with feathers and soft tissues, are almost impossibly unlikely, and allow a tiny window into a world we had no idea existed.

We now also have an incredible evolutionary sequence of fossils from dinosaurs to birds, one of the best for any group of vertebrates. It shows us not only how feathers evolved from very simple structures through to complex flight feathers, but also how flight itself might have developed.

Sinosauropteryx's feathers were a downy fuzz partly used for insulation. This is probably the purpose they served on the earliest feathered dinosaurs too, but over time they began to be used for display, too. The massive *Gigantoraptor* from the Gobi Desert of Inner Mongolia, for example, an 8-metre-long, parrot-beaked oviraptorid, likely used great fans of tail feathers for mating displays. Only later on did feathers begin to be used for flight, as they were in animals such as the four-winged flyers *Microraptor*, *Anchiornis* and *Xiaotingia*. For much of their early history, feathers were likely not used for flight at all. It's difficult to say for certain, however, because by the time the earliest feathered dinosaurs we know of appear in the fossil record, about 160 million years ago, feathers were already being used for all of these purposes.

Four-winged flying dinosaurs have continued to be found. The first, pigeon-sized *Microraptor*, was discovered in 2000. Eagle-sized *Changyuraptor* came along in 2014, and there are likely others yet to be announced. These animals would have had wings on their hind limbs and forelimbs to help them glide between the trees of China's Jurassic and Cretaceous forests. We don't think these animals were capable of the full powered flight of modern birds, as they lacked the keeled sternum to which large flight muscles are anchored in birds today.

The four-winged model was one early experiment in flight within the group of animals from which birds evolved, and many early birds would have had large feathers on their hind limbs and feet too, as some chickens, most obviously bantams, do today. But, as evidenced by *Yi qi*, we're also starting to realise that there were other weird experiments in flight going on during the Jurassic.

When feathered dinosaurs started to be found, some experts felt it was unlikely that the largest carnivores, such as *T. rex*, had feathers. But a series of finds have turned this idea on its head. A number of tyrannosaur species was found with feathers in China including *Dilong* in 2004, and the much larger *Yutyrannus* in 2012. This shaggy, 9-metre-long creature was much nearer in size to *T. rex*, making it much more plausible that the tyrant king itself had feathers.

Other feathered dinosaurs have been really weird animals, such as the herbivorous therizinosaur *Beipiaosaurus*, which had huge claws and a pot belly that possibly acted as a fermentation chamber for vegetation. Another odd one is *Sciurumimus*,[8] the 'squirrel mimic', from Germany, which had a great big bushy tail of fluffy feathers. There's also *Kulindadromeus* from Siberia, described in 2014 (see chapter 6). This one was interesting because it's a little herbivorous ornithischian dinosaur distantly related to the carnivorous theropods and yet it has feathers too, which suggests that feathers may have been spread much more widely across the dinosaurs than most experts supposed. Some palaeoillustrators have already started to draw larger ornithischians, such as the ceratopsian *Pachyrhinosaurus*, with feathers. Only time and more discoveries will tell if these reconstructions are realistic.[9]

China's golden age of discovery

Several years ago I walked into the hangar-like museum space and dig site at the World Dinosaur Valley in Lufeng, 60 kilometres west of Yunnan Province's capital, Kunming. It is an experience I will never forget. The huge public gallery was cool, dimly lit and had an earthy smell about it, but once my eyes had adjusted from the brightness outside I realised this place was very special indeed. All around me were large dinosaur specimens, more than I had ever seen together in one place. These were not models or casts, but carefully reconstructed, largely complete fossil skeletons of nearly 70 Early Jurassic dinosaurs. They were arranged in groups and lit atmospherically from below, so some seemed to float above me as I walked through the columns beneath them.

The oldest specimens are from the Early Jurassic, making the site one of the oldest dinosaur deposits in the world. Most of the specimens are those of primitive prosauropods, such as 9-metre-long *Lufengosaurus*, dated from about 190 million years ago. Like similar prosauropods such as *Glacialisaurus* (see chapter 11) found on Antarctica's Mt Kirkpatrick and *Massospondylus* from South Africa, these animals are related to the ancestors of giant sauropods that later reached lengths of nearly 40 metres. Prosauropods did not reach the same titanic proportions, but already evolution had begun to elongate their necks and sculpt them into the more familiar sauropod body plan. And although these prosauropods spent some time on all fours, they were still partly bipedal, standing on their hind limbs to reach into the trees for fresh vegetation.

It's exciting when museums have single real fossils on display rather than plaster casts, but when those fossils are largely complete and there are nearly 70 of them, it's little short of astounding.

At Lufeng, more than 120 specimens of around 40 species have been found. Ten of these skeletons have more than 90 per cent of their bones, another 60 have more than 70 per cent. Nearly twice as many dinosaurs have been found at this single site in China than have been found in many other entire countries, Australia included.

'This is a classical site for the study of Chinese dinosaurs. It's one of the earliest sites where Chinese palaeontologists recognised fossils and started working on them in the 1930s', says World Dinosaur Valley's director Wang Tao. The Lufeng dinosaurs are also important because they fill a gap in our knowledge of Early Jurassic dinosaurs, which are poorly represented as fossils in other parts of the world.

Little known outside China, museums such as that at Lufeng in Yunnan, at Zigong in Sichuan and at Pingyi in Shandong have some of the largest and most impressive dinosaur collections and display spaces in the world. Public interest in fossils is now increasing and infrastructure is getting better, so regional governments have the money to build even more new museums and hire palaeontologists.

Remarkably, the rate of dinosaur discovery in China could increase even further; many regions have rocks of the right age for dinosaurs but are yet to be thoroughly explored for fossils. The diversity of dinosaurs coming from China is unparalleled, the species ranging from *Epidexipteryx*, which could sit in the palm of your hand, to bus-sized *Yutyrannus*, and much larger new sauropods. 'You have lots of potential in China to find different kinds of dinosaurs, some really big, some really small, some previously unknown from this region', Xu says. 'In terms of discoveries, we are still in the golden age. Many different research groups are all very active, from southern China to northern China, and [working on rocks] from the Early Jurassic to the Late Cretaceous.'

Another region of China where Xu himself has been running regular IVPP expeditions is Xinjiang in the far north-west, the province of his birth. Over 20 years the digs have resulted in the discovery of more than 100 dinosaur fossils, including 10–20 ceratopsians or horned dinosaurs, many of which are new to science. In 2006 they found a primitive Jurassic ceratopsian they called *Yinlong*, which walked on two legs and had a small head crest. It would have been just over a metre long, weighed about 15 kilograms and was dated to about 158 million years old, making it the earliest known member of the group.

In late 2015 they revealed a related species, *Hualianceratops*, from the same deposits. Both these animals were around much earlier than the ceratopsians with which we're more familiar – larger horned dinosaurs with elaborate headgear, from the Late Cretaceous of North America, such as *Triceratops*, *Regaliceratops*, *Styracosaurus* and *Spinops* (see chapter 4). 'About 160 million years ago the ceratopsians had already appeared in western China and they are kind of strange', Xu says. These animals seem to have a mix of features of the horned dinosaurs and also of the bipedal, dome-headed dinosaurs related to *Pachycephalosaurus*. 'Some people believe they are not really ceratopsians and are a basal, more primitive, group ancestral to both ceratopsians and pachycephalosaurs.' When people talk about ceratopsian dinosaurs, they are usually thinking of animals from the Late Cretaceous, but these Chinese discoveries highlight a potentially whole new world of early horned dinosaurs in earlier, Jurassic sediments. In general, the Mid- to Late Jurassic is a period not well documented in many places around the world.

Fifteen years ago, most of the fossil-hunting expeditions in China and any large-scale excavations were organised by the IVPP in Beijing, but today there are many groups of palaeontologists finding fossils in provinces such as Gansu, Henan, Zhejiang and Shanxi. This is in addition to all the fossils from Liaoning, Inner Mongolia, Yunnan and Sichuan, where there has been a longer tradition of dinosaur research.

When Xu was a student in the early 1990s, he felt like he knew all of the dinosaur research that was going on in China and he knew when new discoveries would be announced. But now he can't keep up with everything, and when he travels to other parts of China, he constantly learns about new species about to be described.

China's feathered dinosaurs are now world famous, and Xu himself has become a celebrity figure in China, which accounts for some of the increased interest in palaeontology there. But the dramatic increase in research groups in other regions of China has also been because the IVPP is filling up, says Sullivan, so when students there graduate they often have to go elsewhere if they want to continue their careers in vertebrate palaeontology.

He adds that other parts of the world, such as Alberta, where he has done some work recently, have incredibly rich fossils beds, but people have been hunting in them since the beginning of the 20th century. In China, the fossil beds are equally rich, but many of them are only just being explored. 'There's a bit of a gold-rush atmosphere', Sullivan adds. 'It's a lot of fun.'

Mechanism of flight

It's probable that gliding membranes were present in all the scansoriopterygid dinosaurs, and this explains their puzzlingly long fingers. 'I tend to believe that *Epidexipteryx* and *Epidendrosaurus* both have wings like *Yi*', says Xu. 'All the members of this group probably have styliform elements and membranes attached. You need fossils to support this, but if you ask me I believe it's possible.'

It's not clear what kind of flight mechanism these 'bat' dinosaurs employed, but they may have used a mixture of gliding and flapping. Xu's team attempted, without much luck, to make structural models based on the fossil, to test them aerodynamically. Now they are creating three-dimensional computer models instead. The experts also reappraised the handful of other scansoriopterygid fossils in light of what they now know about *Yi*, but haven't yet found evidence of the styliform element or membranous wings.

'Given that the scansoriopterygids are so interesting and so bizarre, we want to do more work on this group', Xu says. He plans to search for their fossils in several localities in Inner Mongolia and Hebei. 'I don't know how good the possibility is, but we'll definitely try to find more fossils, hopefully some that are more complete and better preserved, fossils that will give a better idea of what these animals looked like.' He has a rough idea of the quarries *Yi* and *Epidexipteryx* came from, and plans to return to both locations.

Sullivan agrees that finding more fossils is an exciting prospect, and he wants to cast the net even wider. 'One of my fantasies as a palaeontologist is that scansoriopterygid material will turn up in the Jurassic of another region of the world, in an area that has very different preservation', he says. Fossils from north-eastern China are often very complete but are compressed near flat, so it can be difficult to see the original shape of the bones.

The really big question is why this group had evolved a second method of dinosaur flight when many closely related theropod lineages had species with very large flight feathers. These include animals such as *Changyuraptor*, an eagle-sized, four-winged flyer described in 2014[10] and *Zhenyuanlong*, a 2-metre-long dromaeosaur related to *Velociraptor*. *Zhenyuanlong*, found in the 125-million-year-old deposits of the Yixian Formation in Liaoning, has the largest feathered wings yet found on a dinosaur. Revealed by palaeontologists Lü Junchang and Stephen Brusatte in 2015, it was described by Brusatte as a 'fluffy, feathered poodle from Hell'.[11] *Zhenyuanlong* probably couldn't fly because it was too large and, relatively speaking, its wings too small, but it may have evolved from dinosaurs similar to *Microraptor* and *Changyuraptor* that could.

'Why evolve a completely different flight mechanism and body plan?' asks Xu. 'This is really bizarre and so far I don't have a good answer.' He believes that perhaps whenever big evolutionary transitions take place – such as that from terrestrial theropod dinosaurs to flying birds – you get strange experiments happening on the fringes, maybe because of pressure from the environment, or because evolutionary constraints have been relaxed.

A hidden world of dinosaurs

The discovery of *Yi* tells us how little we really know about the diversity of dinosaurs – and suggests it's very likely that we will find even more bizarre species in the future. We may only have discovered a fraction of the true diversity of dinosaurs.

In any modern ecosystem, there are species that are very common, but often alongside them, says Xu, are a number of bizarre animals – much rarer species, represented by relatively few individuals. Because of the way fossilisation works, rare species are very unlikely to end up being preserved in the rocks (Bill Bryson, for example, suggests that the chances of a bone being fossilised are about one in a billion),[12] and consequently we are very unlikely ever to know that these species existed. The same goes for animals that are less likely to fossilise for other reasons – they live in mountains where sediments are not laid down, or rainforests where corpses rapidly decompose, or they are juveniles with softer, more fragile bones.

In the modern world, humans have taken over so much of the planet that the majority of wild animals have already effectively vanished from the fossil record of the future. Gaia Vince, author of *Adventures in the Anthropocene*, estimates that if you weighed all of the earth's land vertebrates, humans alone would make up a third of the biomass, while domestic animals and other species that live alongside humans – such as cows, sheep, goats and rats – would make up most of the remainder.[13] Only 3 per cent of the weight would be wild animals, such as elephants, bears, wolves, wombats, armadillos and orangutans. Ten million years from now, fossil hunters will see a very clear mass extinction marking the beginning of the Anthropocene epoch, where the fossils of a great diversity of animals are replaced by those of cows, chickens, humans, cars, skyscrapers, aluminium cans, plastic bags and smartphones.

Rare and unusual species of dinosaur would also have been unlikely to fossilise. 'Over the last 200 years we have found many different dinosaurs and dinosaur groups', Xu says, 'but my feeling is that what we have found so far are the common kinds of dinosaur, but there are many, many rare kinds of dinosaur. It is less likely that they are preserved as fossils and less likely they will be found by palaeontologists. But if we are lucky enough then probably we will find some of those species'.

One idea he has to better estimate the true number of dinosaur species is to look at modern ecosystems and see what proportion of species is relatively common or rare, and use that as a proxy to predict what proportion of dinosaur species we might have missed. Furthermore, he believes that feathers might have been a factor intimately tied to the diversity of dinosaurs. Not only do feathers protect and insulate birds and allow them to fly, but they also increase the rate at which new species form, by acting as so-called isolation mechanisms, through distinguishing colours, for example, which genetically separate populations from one another, meaning they no longer interbreed. 'Many bird species are nearly identical in their skeletal system, but they have different-coloured feathers, and that's what makes them different species', Xu says. This is the main reason birds are a very numerous group today, with around 10,000 living species. 'If many dinosaurs are feathered then they can change their colour relatively easily, so the diversity should be very high if you use birds as a model', he says.

In recent years, the feather colours have been worked out for a series of feathered dinosaurs and early birds – including *Sinosauropteryx* (ginger and white), *Microraptor* (blue–black and iridescent), *Anchiornis* (black and white with a red head crest) and *Archaeopteryx* (black and white) – using electron microscopes to examine the structure of tiny packages of pigment called melanosomes in very well-preserved feather fossils (see chapter 6). Xu believes that the study of colour and melanosomes in China's feathered dinosaurs will be a much bigger area of focus in the future, as it has only been applied to relatively few species so far. One thing he would like to test is whether fossils we currently class as a single species based on skeletons might have been different species with very different-looking feathers. If he has any luck with this technique it could tell us whether or not we are drastically underestimating the diversity of dinosaurs based on bones alone.

Another area of research related to fossil feathers is studying how the colours and patterns of feathers changed as dinosaurs grew and developed. Changes in the types of feathers during development have already been shown in the ostrich-like dinosaur *Ornithomimus* from Alberta in Canada. Darla Zelenitsky from the University of Calgary and François Therrien from the Royal Tyrrell Museum of Palaeontology found different patterns of plumage in juvenile and adult specimens preserved in sandstone. 'This dinosaur was covered in down-like feathers throughout life, but only older individuals developed larger feathers on the arms, forming wing-like structures', Zelenitsky told reporters. 'This pattern differs from that seen in birds, where the wings generally develop very young, soon after hatching.'[14]

The discovery was also interesting because it was the first feathered dinosaur from Canada and also the first not recovered from the very fine-grained sediments of China and Germany. Before this, scientists thought feathered dinosaurs were only likely to be preserved in muddy, possibly volcanic sediments deposited in calm lakes and lagoons. As most fossils are preserved in sandstone, the discovery suggested there is much greater potential for finding feathered dinosaurs in other parts of the world.

Already, Xu's team have started to study the melanosomes on a limited number of places of the fossil of *Yi qi*, giving them a rough idea of the colour of this bat-like dinosaur. 'Based on our basic understanding of colour and melanosome morphology, I will say *Yi qi* probably had a green or brown colour', Xu says. 'It's a guess, but the probability is high because the melanosomes appear similar to modern birds with green or brown feathers.' If proven correct, this is the first time that green has been detected in a feathered dinosaur fossil.

The weird wonders that will follow

The majority of China's feathered dinosaurs have come from western Liaoning's Yixian and Jiufotang formations – layers of rock containing a single evolving community of plants and animals known as the Jehol Biota. This dates from the Early Cretaceous period 120–125 million years ago.

Yi qi instead comes from the Yanliao or Daohugou Biota and the Tiaojishan Formation of rocks from Hebei, Liaoning and Inner Mongolia. The type of preservation and incredible fossils are similar to those of Jehol, but the rocks here are about 30 million years older, from around 153–165 million years ago during the Mid- to Late Jurassic.[15] These rocks were also the source of *Anchiornis*, *Xiaotingia*, the other scansoriopterygids *Epidexipteryx* and *Epidendrosaurus*; a few other very bird-like feathered dinosaurs; and also pterosaurs, amphibians, lizards and mammals. Furthermore, because of the geographic overlap with Jehol and the fact that the fossils here are tens of millions of years older, palaeontologists have a very unusual opportunity to study changes in the make-up of the community over time and during a key period of dinosaur and bird evolution.

And it's very likely that exciting new finds are yet to come from the Yanliao Biota. Sullivan is betting on a string of strange species in the next few decades. 'Some of them I would expect to be every bit as weird as *Yi qi*. We're seeing a side of the Jurassic we've never had access to before, and I'm sure there are going to be a lot of surprises.'

Xu agrees that Hebei and the surrounding regions will be a productive source of strange new species. '*Yi qi* was totally unexpected. We couldn't believe it. If you know dinosaurs very well and the transition well, then you'd never expect there would be a dinosaur with bat-like or pterosaur-like wings instead of feathered wings. Discoveries like this will continue to emerge and demonstrate how complex the transition to birds was', he says. 'I would not be surprised if we find even more bizarre species in the future.'

Palaeontology is always an adventure, no matter where it happens, but in the early twentieth century, one man took fossil hunting to swashbuckling extremes.

3

DWARF DINOSAURS AND TRAILBLAZING ARISTOCRATS

TRANSYLVANIA, ROMANIA

> *Hațeg (pronounced Hat-zeg) was a strange place during the Late Cretaceous: an island inhabited by peculiar dwarf dinosaurs. The first of these was discovered by a trailblazing Transylvanian aristocrat – a motorcycling World War I spy who almost became the king of Albania.*

Seventy million years ago, as the Cretaceous Period nears its conclusion, global sea levels are high and Europe is an island archipelago. In the east, in the area that in millions of years will be the Romanian region of Transylvania, is a large island lapped by warm tropical waters and long isolated from the mainland. Here live strange, primitive dinosaurs and pterosaurs, which are quite unlike their contemporaries in other parts of the world – and in contrast to Transylvania's mythological vampires, these sometimes bloodthirsty biological oddities really did exist.

Circling on thermals hundreds of metres up is one of the largest creatures that has ever flown; she has a skull 3 metres long, and when she stands tall on the ground she is the height of a large bull giraffe. Now, though – having unfolded her great leathery wings with a loud whip-like crack and catapulted herself into the air — she is the size of a small biplane.

This pterosaur, a giant flying relative of the dinosaurs, is covered in a pelage of glossy, fur-like filaments, which gently ripple in the wind as she thunders past, sucking vortices of wind around her wings. Like an eagle today, this *Hatzegopteryx* has the keen eyesight of a predator, and spots below her favoured quarry. Her resolve determined, she wheels around with her great 12-metre wingspan and swoops towards the ground. Landing with a resounding *thwump*, she folds in her wings and awkwardly struts on stilt-like limbs towards a juvenile long-necked sauropod dinosaur. She claps the hapless creature in her huge jaws, shakes it, tosses it into the air and swallows it in a couple of gulps.

Across the planet sauropods are king — titanic herbivores up to 80 tonnes and 35 metres long — but not here on Hațeg Island. Researchers confirmed in 2010 that these dinosaurs had been dwarfs, no bigger than the size of a cow or a horse. Evolution on islands can do strange things to the physiology of animals, shrinking or supersizing them — or taking away their ability to fly, as seen with Pacific birds today.

Hatzegopteryx (described in a *Wired* magazine article as a '16-foot-tall reptilian stork that delivered death instead of babies')[1] was named in 2002 by French palaeontologist Eric Buffetaut and his Romanian colleagues Dan Grigorescu and Zoltán Csiki. Research in the past decade has uncovered much about the way it lived, and some estimates suggest that the giant azhdarchid family of pterosaurs, such as Hatzegopteryx and its Texan relative Quetzalcoatlus, might have been able to fly 16 000 kilometres without stopping, and stay aloft for a week at a time or more.

The snack-sized sauropods, the juveniles of which Hatzegopteryx may have fed upon, are the smallest ever discovered. Magyarosaurus dacus, at 6–7 metres long and weighing no more than a tonne, was a pygmy compared to South American titanosaurs. Some of those, such as Dreadnoughtus or Argentinosaurus, were 100 times its size. Other Haţeg species, such as the duck-billed hadrosaur Telmatosaurus transsylvanicus, were also very small.

A century of fossil collecting at Haţeg has yielded 10 specimens of Magyarosaurus. This is enough to be sure the small size wasn't a freak condition among a few anomalous animals, and in 2010 research led by Professor Mike Benton at the University of Bristol used the microstructure of the bones to confirm that the animals had been adults and not juveniles.

These really were strange dwarf dinosaurs, as had first been predicted a century before by visionary man of science and Transylvanian aristocrat Franz Baron Nopcsa von Felső-Szilvás, who found the first of these dinosaurs on his family estate in the 1890s. Nopcsa rode a motorbike, was a spy in various Balkan states in the years preceding and during World War I, was briefly considered for the throne of the nascent kingdom of Albania, and eventually killed himself and his long-time secretary and boyfriend Bajazid Doda in double murder–suicide in 1933, just a few months after Adolf Hitler swept to power as chancellor of Germany.

Nopcsa was conspicuously ahead of his time, and theorised many concepts we take for granted today: he deduced that birds evolved from dinosaurs, that some dinosaurs and pterosaurs were warm-blooded, that Transylvania was then an island, and even that plate tectonics explained puzzling patterns in the fossil record. Belgian palaeobiologist Louis Dollo described him as a 'a comet racing across our palaeontological skies, spreading but a diffuse sort of light'.[2] Nopcsa was a legendary dinosaur hunter on a par with the late-1800s US palaeontologists Edward Drinker Cope and Othniel Charles March, of 'Bone Wars' fame.[3] Despite this, he was derided by some of his contemporaries, and remains very little known today.

Trailblazer and troublemaker

Nopcsa is one of the most colourful figures in the whole history of palaeontology, and fantastic anecdotes about him abound – too many to recount them all here. British palaeontologist Gareth Dyke, who is writing a biography of Nopcsa, recounted one of my favourites in a 2011 article for *Scientific American* magazine.

The year is 1906. Nopcsa is in London's shiny new Natural History Museum (then the British Museum, Natural History). He approaches the giant cast of *Diplodocus* in the entrance hall, the one that remained there until 2015 but at that time had very recently been installed. Nopcsa leans over, carefully removes one of the large toe bones from its mount, turns it over, then slips it back into place the correct way up. According to the baron's later writings, museum officials were not amused.

Many anecdotes about Nopcsa involve how he set out to cause controversy or have a discreet laugh at the expense of the establishment. One such tale is about the description of a species of Cretaceous fossil turtle, which he named *Kallokibotion bajazidi*. 'The genus is *Kallokibotion*, the species is *bajazidi* after his lover Bajazid', says Dyke from Hungary, when I speak to him over the phone. He tells me Nopcsa was a bit of a troublemaker: 'The name *Kallokibotion* literally means "beautiful box". The name means "beautiful box of Bajazid". He named the turtle after his lover, because the shape of the shell reminded him of Bajazid's arse'.

Nopcsa met Bajazid Elmaz Doda, a Macedonian who would henceforth be described as his secretary, in 1906, and, according to Robert Elsie a UK-based historian of Albania, 'hired the eighteen-year-old lad as his servant. This relationship evolved into a love affair and a long-term domestic partnership'.[4] Bajazid was a geologist and talented fossil hunter in his own right – he discovered the bones of an armoured ankylosaur dinosaur, which Nopcsa named *Struthiosaurus transylvanicus* in a 1915 monograph. In his own memoirs, Nopcsa would later recount: 'On 20 November 1906, in Bucharest, I met Bajazid Doda. Since that time, Bajazid has been with me and ... he has been the only person who has truly loved me and in whom I had full confidence, never doubting for a moment that he would misuse my trust'.

Dyke says that aside from being a troublemaker, Nopcsa was also arrogant and competitive. 'There are stories about him hearing that somebody had climbed a mountain in the Carpathians and then immediately going to climb a higher one, just so that he could be more successful, more manly than anyone else.'

Nopcsa's palaeontological story begins in 1895, when he was 18 and his younger sister, Ilona, found some puzzling fossil bones on the family estate at Szacsal, in the foothills of the Carpathian Mountains, then part of the Austro-Hungarian Empire. Nopcsa took the bones to Vienna, where he was studying at the university, and showed them to geology professor Eduard Suess. Following Suess's advice, he took it upon himself to read up on anatomy and geology. Within a few years he'd written a paper describing Transylvania's first dinosaur as *Limnosaurus* (the small duck-bill now called *Telmatosaurus*).

Nopcsa's wealth, brilliance and connections might have secured him any career he chose. 'Yet Ilona's serendipitous discovery at Szacsal provided the spark for Franz's choice of career, and he committed himself thereafter to the study of these fossils', says Professor David Weishampel.[5] This was no mean feat, as at the time the university had no vertebrate palaeontologists. Nopcsa read widely, corresponded with foreign experts, and taught himself anatomy and physiology. Time spent with his head down in the laboratories and libraries of Vienna was complemented by summers in the field, prospecting for dinosaur fossils in and around Haţeg. Men on his family estate also began to bring him bones from eroded rocks that had appeared in the annual spring snowmelt.

In July 1899, when he was 22, he gave his first lecture at the Vienna Academy of Sciences on the 'dinosaur remains from Transylvania'. In the following years he described the region's Sânpetru sandstone as a stratigraphic unit, and started work on his description of the Haţeg dinosaurs in the first of five monographs. All up, over the next 35 years, he would publish more than 100 papers on his fossil studies (and nearly 200 publications in all). He travelled widely across the mountains of northern Albania for 15 years, writing the first detailed accounts of their geology and geography. Later, this knowledge would provide support for the concept of continental drift.

'Not all his theories were accepted at the time, but they did succeed in advancing and stimulating a wide range of fields of palaeontology', says Elsie.[6] 'Equally important were Nopcsa's achievements in the field of geology, an example of which was his research into the tectonic structures of the western Balkan mountain ranges.'

Banditry and espionage

Nopcsa, became utterly fascinated with Albania early on in his life, and he would devote as much time to the study of the geology, topography, geography and people of Albania as he did to Transylvanian dinosaurs.

In his memoirs – which are full of encounters like this – the baron recounts how for several weeks in 1907 he was held hostage in the Dibra Mountains of Albania by a notorious bandit, writing: 'Neither Bajazid nor I appreciated all the excuses and we began to suspect that something was amiss. In the afternoon, Mustafa Lita called for Bajazid and informed him that we were his prisoners. He was demanding ten thousand Turkish pounds for our release'.[7] In the end, Bajazid's father turned up with 'ten armed retainers' to rescue them and all was right again.

Nopcsa lived somewhat of a double life, flitting back and forth between time spent dressed in rural clothes, exploring the back country of Albania on horseback, and periods dressed in nobleman's finery at the family estate in Szacsal. After a visit from Arthur Smith Woodward – then head of palaeontology at the British Museum, Natural History (see chapter 4) – and his wife in 1906, Lady Smith Woodward noted some of the customs at the Nopcsa family castle. She described how the clothes of the local peasants contrasted with the extravagant dress of Franz Nopcsa, who was decked out in a 'gold brocade tunic and a dolman lined with sable skins slung over one shoulder, white buckskin breeches, high shining black boots with gold embroidered tops, sword belt, sword, tunic buttons, all gleaming with glittering gems, and a fur cap with a high egret plume'.[8]

In his 1945 biography of Nopcsa, Andras Tasnádi Kubacska writes that the baron was constantly restless and 'dissatisfied in the very heart of his being':

> Only unwillingly and reluctantly did he follow the beaten path of the great multitudes. He was just as seldom exclusively a scientist, or only an explorer, as other times only a spy or politician. When he roamed through the gorges of the Balkans, his unsettled blood drove him from one adventure to the next ... There was hardly a museum between Budapest and London in which he had not worked: Vienna, Munich, Stuttgart, Tübingen, Frankfurt am Main, Berlin, Basel, Zürich, Brussels, Paris, Le Havre, Bordeaux, Marseilles, Cambridge, Oxford, London, Bologna, Bucureşti [Bucharest], Moscow and so on were the destinations of his

scientific travels. When he needed relaxation, he then threw himself into the arms of politics, but when he began to hate everything, he went to the Carpathians to hunt bears and sheep ... He often vanished for a long time, but then he would appear completely unexpectedly and enjoy hospitality for days, weeks, indeed often months at a time. Nevertheless, one beautiful day he would go away again and perhaps newly appear for the first time several years later.[9]

Albania was part of the Ottoman Empire until the early 20th century, but in the run-up to the Balkan Wars of 1912–13, which saw the Turks lose all their European territories – Nopcsa was among those pushing for it to become an independent nation linked to the Austro-Hungarian Empire. 'He wanted to establish Albania as a nation, with a capital and borders and everything else', says Weishampel, a palaeontologist at Johns Hopkins University in Baltimore. 'The first and second Balkan Wars involved the eventual demise of the Ottoman Empire and the creation of Albania. But it was a violent birth, because the Soviets and the Ottomans didn't want that, and there was Austria-Hungary sitting in the background eager to have a connection to the Mediterranean in the Adriatic. They found out that Nopcsa was making maps – and if you're making maps and you're working on your own, you're a pretty good candidate for espionage.' So, he was coopted to be a spy for Austria-Hungary during the Balkan Wars.

But Nopcsa's plans for his own involvement in local geopolitics were much grander. With his passion for all things Albanian, and his aristocratic heritage, he wanted to become king of this nascent nation, and he used his connections to raise the idea at the highest political levels. His plan was to arm the tribes of northern Albania to start a guerrilla war against the Turks, Weishampel says. 'A successful routing would result in Albanian independence, with Nopcsa himself as king under the protectorate of Austria-Hungary.' His strategy – which he outlined in a telegram sent to the Austro-Hungarian chief-of-staff – was to go to New York City, marry a rich American and spend the money on ships and guns with which to arm guerrilla warriors to defeat the Turks. He would pledge his allegiance to Austria-Hungary, and in exchange it would recognise Albania as a nation with Nopcsa on its throne.

But it wasn't to be. As far as we can tell, Nopcsa never received a reply to his telegram. His influence in the Balkan region waned following the murder in 1914 of Archduke Franz Ferdinand, the heir to the throne of Austria-Hungary – the event that launched World War I and plunged Europe into the greatest conflict it had ever known.

During the war, Nopcsa continued to work as a spy, this time in western Romania and for the prime minister of Hungary – a role for which he was uniquely prepared, after his many years of solo exploration in the region. Kubacska writes about one occasion in 1916, when Nopcsa was sent by the Albanian defence minister on a secret mission to Romania:

> Disguised as a shepherd, he made his way before daybreak to reach the border hut on the ridge of the Carpathians ... He struggled to the summit only with great difficulty, but lost direction in the mist and snow flurries on the descent. He wandered for many hours, having lost the path and without directions ... He had not the slightest idea where he was or where his feet were carrying him. Bad weather raged all around him and soon the snow whipped at him from this and that direction. His footprints blew away in an instant and Nopcsa knew neither forwards nor backwards.

Nopcsa apparently survived that misadventure – and many other dramatic encounters – through sheer force of will. But when the Central Powers, including Austria-Hungary, lost the war in 1918, it was the end of Nopcsa's career as a spy.

It was the end, too, of his family estate in Transylvania, because Transylvania was ceded to Romania and his family's property was seized by the state. 'Every bloom, every bush and every footstep on the ground carried the memories of beautiful times', he later wrote. 'It was if my entire past vanished into a chasm before my eyes.' It was a turn of events from which the baron would never recover.

Rich dinosaur-hunting ground

Though today it's a town in Romania hundreds of kilometres inland, 70 million years ago Hațeg was an 80 000-square-kilometre island about the size of Tasmania. Part of a southern Europe that was much warmer than today, it was flooded by shallow seas and was home to dinosaurs, as well as diverse communities of tropical plants, insects, fish, frogs, lizards, birds and mammals, which are also recorded in the fossil record. We now know of about 14 species of dinosaur from Hațeg.

These fossils are notable for a number of reasons – they are some of the most recent dinosaur remains from Europe, from a time right at the end of the Cretaceous when much of Europe was under the Chalk Seas that gave rise to England's White Cliffs of Dover. Hațeg is also one of those rare localities where the skeletons of adult dinosaurs have been found in association with nests of eggs, although there is debate about whether the eggs are those of sauropods or hadrosaurs.

One of Nopcsa's revolutionary ideas was that the dinosaurs of Hațeg represented island-dwelling species, which he believed accounted for their miniature stature and the fact that they seemed primitive compared to their contemporaries from the mainland. Three kinds of Hațeg dinosaur – the herbivorous ornithopods *Telmatosaurus* and *Zalmoxes*, and the sauropod *Magyarosaurus* – were much smaller than related species from the United Kingdom, Germany and North America. *Telmatosaurus* ('marsh lizard'), the first dinosaur described by Nopcsa, was a hadrosaur that grew to about 5 metres long – significantly smaller than its 9-metre-long mainland relatives.

Nopcsa then described the sauropod now called *Magyarosaurus*, choosing the species name *dacus* in reference to an ancient tribe of Romania. At the time, however, he named the animal *Titanosaurus dacus*, perhaps as a wry joke of some kind, given Transylvania was then firmly part of Hungary (*Titanosaurus* also seems inappropriate given the size of the beast, but perhaps this was ironic too).

In any case, the name was officially changed to *Magyarosaurus dacus* in 1932 by German palaeontologist (and Nopcsa's friend) Friedrich von Huene at the University of Tübingen, because the name *Titanosaurus* had already been applied to another dinosaur before Nopcsa used it in 1915. Von Huene's name refers to the ancient Magyar people of Hungary and may have aimed to set the record straight.

Nopcsa was also one of the first to suggest differences in the body size and shape of male and female dinosaurs (something biologists call sexual dimorphism). He speculated that some herbivorous dinosaurs showed sexual differences in the fossils, such as in the robustness of the pelvis and the presence of bony ridges on the skull of some specimens (he also suggested the presence of an attachment point for penis musculature on the pelvis, but this was largely ignored).[10]

In a 1914 paper about the Transylvanian finds, Nopcsa noted that while the turtles and crocodilians 'reached their normal size, the dinosaurs almost always remain below their normal size'. Taking an ideological leap, he proposed that these animals were dwarf species rather than simply juveniles, and that Haţeg had been an island. It wasn't until the 1980s and 1990s that geological evidence, such as the surrounding marine deposits, confirmed that Haţeg really had been an island 200–300 kilometres from the nearest major landmass.

Nopcsa's ideas are also borne out by what we know about the

evolution and survival of island-dwelling species today. The prehistoric elephants of the Greek island of Crete were dwarfs, for example, and the *Homo floresiensis* 'Hobbit' people of the island of Flores in Indonesia were dwarfed relatives of our own species (Flores also had pygmy elephants and supersized Komodo dragons). We also know that some 'primitive' species that had already become extinct across much of their original ranges were able to hang on much later on islands – such as the mammoths of Wrangel Island in the Arctic, where a population of 500–1000 continued to survive until around 3600 years ago, despite the species having disappeared across mainland Eurasia about 6000 years earlier (in other words, there were still mammoths in 1500 BC, halfway through the reign of the Egyptian pharaohs, when the pyramids of Giza had already been standing for 1000 years).[11]

'The idea of "island dwarfing" is well established for more recent cases', Mike Benton told reporters in 2010. 'There were dwarf elephants on many of the Mediterranean islands during the past tens of thousands of years. These well-studied examples suggest dwarfing can happen quite quickly. The general idea is that larger animals that find themselves isolated on an island either become extinct because there isn't enough space for a reasonably sized population to survive, or they adapt. One way to adapt is to become smaller, generation by generation.'

Though Nopcsa's ideas were widely accepted, it wasn't until 2010 that researchers led by Benton and Zoltán Csiki re-examined the fine internal details of Haţeg's fossil bones to show beyond doubt that that *Magyarosaurus*, *Telmatosaurus* and *Zalmoxes* had been adults and had all undergone a form of accelerated development, whereby they became sexually mature at much smaller sizes and younger ages than their mainland relatives.

'They had been claimed as dwarfs ever since Nopcsa found them', Benton tells me when I meet him in Bristol, 'but he had not provided proper numerical or bone histological [staining for microscopy] treatments. We wanted to face the question head-on and demonstrate the claim was right or wrong once and for all. This required geological evidence that Hațeg and other dinosaur spots were islands, that the dinosaurs were all adult and not just juveniles, for us to work out the difference in size from nearest relatives, and to consider whether they were somewhat "primitive" for their time'.

Magyarosaurus and the dinosaurs of Hațeg are not the only Cretaceous dwarfs we know about. In 2006, Professor Martin Sander of the University of Bonn and his team announced the discovery in Germany of a sauropod they called *Europasaurus*.[12] The 11 specimens ranged in length from 2 to 6 metres. They were initially thought to be juveniles, but an analysis of the bone structure revealed that, like *Magyarosaurus*, they were miniature adults. A titanosaur, *Neuquensaurus australis* from South America, is also estimated to have grown only to 7–9 metres. Both these species are likely to have been island dwarfs too.[13]

Dinosaur bones have growth marks, something like the annual rings of trees. In young dinosaurs, when growth is rapid, these have wide spaces between them, but as the animals mature, the marks fall much closer together – and this is what Sander's team found with *Europasaurus* and the Transylvanian dinosaurs by examining thin cross-sections of fossil bone under a microscope. A further analysis of the shape and structure of *Europasaurus* skulls in 2015 backed up the idea that it was a dwarf that reached sexual maturity early.[14]

Mike Benton has studied what appears to be an island fauna with evidence of dwarfing at another Romanian site, too. The Cornet fauna comes from caves near Oradea in the north-west, a very early Cretaceous site that has yielded thousands of bones, mostly of a handful of species of birds, pterosaurs and small herbivorous dinosaurs. The fact that the variety of species is low and that some dinosaurs – such as the ornithopod *Camptosaurus* – are unusually small, was the clue that Oradea had been an island. It would have been sitting in the Tethys Sea at this time (143 million years ago), hundreds of kilometres away from the nearest landmasses in Germany and Russia.

When animals make their way to islands – either by swimming, or by rafting or crossing a land bridge (or their habitat becomes an island through flooding) – they typically find food sources are much scarcer. This creates enormous evolutionary pressure to become smaller, because small animals require less food and have a better chance of survival. This shrinkage can happen remarkably rapidly, sometimes within just 10–20 generations, as observed with deer released onto the Shetland Islands in Scotland. There are two ways dwarfing can occur through development – one is by growing more slowly so that by the time you reach maturity you are still small. The other – which is the kind of dwarfing detected in the German and Transylvanian examples – is by reaching sexual maturity and adulthood much earlier, before growth to a large size has happened.

Weishampel believes that reaching sexual maturity early was the real evolutionary driver of dwarfing on Haţeg. 'Colonising' species that arrive on islands, where resources are scarce, are much more likely to survive if they can reproduce rapidly to increase their population size, and also more swiftly evolve to suit the conditions, he says. The shrinking, therefore, may simply have been a side effect.

Summers in the field

Weishampel has spent many summers in the field in Romania searching for fossils, working with Dan Grigorescu from the University of Bucharest and, in more recent years, with Coralia-Maria Jianu of the Deva Museum of Dacian and Roman Civilization. Here in Transylvania, unlike many parts of the world that yield great fossils, rain is very common. Monthly rainfall can average 13 centimetres and rain lasts for days, sometime sending the palaeontologists to their tents to wait it out. Fossil hunting here can be a challenge too, as most pieces are single bones found in isolation and they are hidden by thick vegetation. A strong back and legs are a prerequisite for clambering up the rocky slopes along the Sibişel River valley.

Many of the things routinely turned up here as small quarries are excavated are the same kinds of bones Nopcsa found a century ago – pieces of *Kallokibotion* turtle shell, *Zalmoxes* and *Magyarosaurus* vertebrae, *Telmatosaurus* shoulder girdles, crocodile teeth – but work led by Grigorescu in the 1980s and 1990s, and now by Csiki, has resulted in many new discoveries, such as the giant pterosaur *Hatzegopteryx* in 2002. In 2010, excavations turned up fossilised scutes – or armour plates from the skin – associated with *Magyarosaurus*. This kind of armour is very unusual in a sauropod, and may have helped protect it against the predatory pterosaurs that prowled the skies and plains of Haţeg Island.

Also in 2010, a new dromaeosaurid was described, which experts believed was related to *Velociraptor*, but it had several puzzling features. *Balaur bondoc* was a raptor unlike any other.[15] It had a weird arrangement of toes on its feet – the forelimbs had just two fingers each, and the hind limbs had not one, but two of the 'hyperextendible' sickle-shaped killing claws that dromaeosaurs use to slash through their prey. The authors of the paper describing the species believed that this unusual anatomy might again have been a strange effect of island life on this predator;[16] but a reanalysis of the bones in 2015 came to a very different conclusion, arguing instead that *Balaur* was a secondarily flightless bird (and therefore still a close relative of the dromaeosaurids).[17] The one specimen is lacking a skull, which would provide a quick answer. With more excavations in Transylvania each year, a skull may yet be found.

Azhdarchid pterosaurs, such as *Hatzegopteryx* and *Quetzalcoatlus*, had wingspans more than three times that of today's largest birds, and have left experts on biomechanics scratching their heads as to how they could have launched themselves skywards, remained airborne and landed without injuring themselves. Recent studies of their bones suggest that they were able to crouch down and use all four limbs to spring upwards with violent force, catapulting – or rather pole-vaulting – themselves into the air.

Pterosaurs were around the entire time dinosaurs were, emerging perhaps 200 million years ago and vanishing in the same extinction event 66 million years ago, but they are not dinosaurs themselves – they are a closely related sister lineage that branched off from a common ancestor. Animals such as *Hatzegopteryx* spent a lot of time on the ground and could walk on all fours, and this may have brought them into contact with Hațeg's dwarf dinosaurs, the young of which may have been a favoured snack. It's also likely that the large size of these predators evolved as a response the lack of large terrestrial predatory dinosaurs on the island (*Balaur*, only found in 2010, is the largest theropod).

And perhaps we would never have known anything of this array of weird Transylvanian creatures if some puzzling bones hadn't been stumbled upon that day in 1895, and Franz Baron Nopcsa von Felső-Szilvás hadn't henceforth devoted himself to revealing their mysteries. Sadly, his life would end in tragedy ...

A man out of time

On the morning of 25 April 1933, Nopcsa awoke in his Vienna home and made tea for Bajazid, into which he poured sleeping powder. Bajazid was an alcoholic by this stage, and we may speculate that he knocked the tea back in his thirsty and hung-over state, thereafter falling into a deep sleep.

Nopcsa retrieved his revolver from his desk and shot his partner in the head, after which he walked into his workroom, wrote a note and then shot himself as well. In Kubacska's reimagining of the scene:

> He gripped the revolver and stuck the barrel into his mouth.
> A muffled crack made the windowpanes lightly resonate.

> The baron sat motionless. His head sank slowly forward, his clothes full of blood. He began to gasp. A slight shudder ran through his limbs, his legs became rigid and the weapon slipped from his now-powerless hand.

Explaining his actions in the note to police, Nopcsa wrote:

> The reason for my suicide is my nervous system, which is at its end. The fact that I killed my long-term friend and secretary, Mr. Bajazid Elmas Doda, in his sleep, without him having an inkling as to what was going on, was because I did not want to leave him behind sick, in misery and in poverty because he could have suffered too much.

The murder-suicide made the front page of the evening newspapers.

Following the loss of his estate at the end of World War I, Nopcsa had worked for a spell at the Geological Institute in Bucharest – the first time he'd had to draw a salary in his life. Later he became the director of the Royal Hungarian Geological Institute in Budapest, where he stayed for two years. But he was autocratic in his rule and not well liked, so it didn't last.

He spent the last part of his life driving around southern Europe on a motorcycle, with Bajazid in the sidecar, looking at things of geological and natural interest and collecting data. In one trip across the Alps and down through Italy, they covered 6000 kilometres. But Nopcsa's money was dwindling, Weishampel says. To compound matters, Bajazid was by now habitually drunk and totally dependent on him. Nopcsa had also suffered a head injury after being attacked in Romania in 1920, which led to recurring bouts of a neurological complaint. The combination of this and his impoverishment led to a deep depression, from which there was this time no recovery.

In his translation of Kubacska's biography, Weishampel notes:

Only occasionally does a person emerge from the background of tradition with the promise of refocusing contemporary thoughts, ideas and attitudes. It is this potential, together with its paradoxes, conflicts, successes and failures, which makes for the truly great arts ... These clashes of ideas, personal lives and worldviews also dominate the greats of science, among them Galileo, Buckminster Fuller, Einstein and Darwin. In no less a way, Franz Baron Nopcsa was to live his life full of scientific breakthroughs and achievements, but in the end lacking in resolution, understanding and peace.

Apart from his discovery of the dwarf dinosaurs, and his work with colleagues and friends on dinosaurs across Europe, Nopcsa also combined his findings on geology and evolution to make significant progress in thinking of fossils as animals that were once part of complex living ecosystems, which was for its time revolutionary. 'Nopcsa often used new and eclectic approaches to solve paleontological and evolutionary problems, always stressing the biological context of extinct vertebrates', adds Weishampel, who says that his innovative approach and financial freedom contributed to Nopcsa's pioneering work in the field of palaeobiology.

As his old tutor Professor Eduard Suess wrote of the baron's untimely passing, 'Those who knew him closely remain under the spell of his personality. The sudden and unexpected end of his enthusiastic and impulsive life fills them with grief. What he has put down in his writings is destined to live on in silence'.

At around the same time that Nopcsa was uncovering the dwarf dinosaurs of Transylvania, fossil hunters halfway round the world in North America were uncovering a series of new relatives of *Triceratops* with curiously elaborate frills and spikes. Little did they know then the great variety of these peculiar ceratopsians still awaiting discovery.

4

HORNY ORNAMENTS AND SEXY CERATOPSIANS

ALBERTA, CANADA

The number of species of Triceratops*-like horned dinosaurs has exploded in the last decade, revealing a remarkable array of elaborate head ornamentation unlike anything seen today. Could it be that sex, rather than violence, was the driver for the evolution of these extraordinarily weird adornments?*

In 2005 geologist and amateur palaeontologist Peter Hews was searching for fossils in the south-west of the Canadian province of Alberta. Here along the Oldman River, in the foothills of the Rocky Mountains, he spotted a triangular fossil sticking out of a cliff on the river's south bank, as well as other shapes that looked like horns. 'It seemed as if I was looking at the snout of a dinosaur', he says, adding that his heart was pounding at this point because he knew how rare and exciting the discovery of a skull was. 'I sent the bone, the horn and some pictures, along with the exact location, to Darren Tanke at the Royal Tyrrell Museum of Palaeontology [RTMP], and his excitement was also palpable. I met him at the site a couple of weeks later and it was determined that this was definitely the front end of a ceratopsian skull, the majority of which was still hidden in the river bank.'

What Hews had stumbled upon was significant indeed – one of the most elaborately ornamented and impressive horned dinosaur discoveries since *Triceratops* itself in Wyoming in 1888. It was the huge 1.5-metre-long skull of a species later dubbed 'Hellboy', because of what the palaeontologists described as the hellish dig to extract it, and the enormous subsequent job of chiselling it from its case of rock in the lab.

The banks of the river where Hews had found the skull are very steep and close to the water's edge. Furthermore, the river is a protected habitat for the endangered bull trout, Alberta's provincial fish, and allowing any rock and sediment from the dig to slip into the water was prohibited. 'You can imagine how, working on such a narrow area, it's very difficult to do that – digging a big hole in the cliff but not having the sediment enter the river', says palaeontologist Caleb Brown, also with the RTMP in Drumheller, Alberta. 'So a coffer dam was built to contain the sediment.'

The rock surrounding the skull was so hard that jackhammers were required to dig it out. The resulting block had to be airlifted out to a truck by a helicopter. All of this took field seasons across two summers to complete. Even back in the museum's fossil prep lab in 2010, the work wasn't over. 'That's not the end of the story. That's just the beginning', says Brown, who came on board after Tanke had completed an 18-month process using specialised methods to free the skull from the rock. But what they found inside was much more spectacular and weirder than they could have guessed, and was well worth the effort and the wait ...

One from the vaults

In 1916, as World War I raged in Western Europe, father-and-son team Charles and Levi Sternberg were excavating a bone bed rich with fossils in the Steveville badlands of Alberta's Dinosaur Provincial Park. They were there to find dinosaurs, and on this occasion were being paid by London's Natural History Museum (NHM) to source attractive new specimens for display. Several decades earlier, Charles had been a fossil collector in Kansas for ED Cope, who was famed for his conflict with OC Marsh in the 'Bone Wars' of the late 19th century.

It was only a decade since the NHM's spectacular replica of a 27-metre long *Diplodocus*, ever since affectionately known as Dippy, had been revealed to the public. This was a gift to Britain's King Edward VII from the Scottish-born millionaire Andrew Carnegie, founder of the Carnegie Institution, which had acquired the original in Wyoming in 1899. Spectacular specimens were big business in the early 20th century, when dinosaurs were experiencing their first flowering in the public consciousness, and species such as *T. rex* had only recently been discovered.

'Museums were really trying to compete to see who could have the best fossil hall', says Andy Farke at the Raymond M Alf Museum of Paleontology in Claremont California. 'There was a rush to have bigger and better displays. The British Museum – now the Natural History Museum – wanted to fill up a hall with dinosaurs, so they hired Charles and his son Levi, who had had great success previously collecting for museums such as the American Museum of Natural History.'

Within the packages of fossils the Sternbergs sent to London (called the Sladen Collection, after a museum donor who funded the expedition) were pieces of the skulls of several ceratopsian dinosaurs. These included the horns and part of the frill, but were missing the upper and lower jaws. Though the remains were scant, Charles believed they might once have belonged to an animal related to *Styracosaurus* – a species that his other son and Levi's brother, also called Charles, had discovered several years earlier at a site a few kilometres away from this new bone bed in Dinosaur Provincial Park.

But when the museum's keeper of geology, Arthur Smith Woodward, eventually examined the fossils, he was greatly disappointed with their quality. A letter he dashed off in irritation, dated 11 January 1918, dismissed Charles's observations about a possible new species, noting that 'there is indeed in the Sladen Collection nothing but rubbish'.[1] Farke says the palaeontologists at the NHM were pretty disappointed, as they'd hoped – based on the Sternbergs' reputation – to get spectacular, complete skeletons. Instead, they had ended up with specimens that looked like rubble.

That was that – and for the next 90 years the fruit of the Sternbergs' labour sat unloved and barely examined in the NHM collections. That was until Farke, an expert on horned dinosaurs, was visiting the South Kensington site in 2004 and, with an hour to spare, decided to browse through some of the palaeontology collection. An unusual specimen caught his eye – it was a piece of a frill of a horned dinosaur with several spikes emerging from it, but it was upside down and buried in rock and plaster. 'I could tell it was a frill, but it didn't look like the frill of any animal I'd ever seen before', Farke says. 'That started getting some alarm bells going off in my head, so I talked with the curator there, Paul Barrett, to see if we could get the fossil cleaned up a little bit.' Subsequent discussion with ceratopsian experts back in North America revealed that this was a legendary specimen some had heard of before – and Phil Currie had once even photographed – but that had since been lost in the bowels of the museum.

When the specimen was prepared, Farke was proved right – it was a previously unknown species with a new arrangement of headgear. With a team including Barrett, Tanke, Michael Ryan and Mark Loewen, Farke published a paper in the journal *Acta Palaeontologica Polonica* in 2011 and named the new species *Spinops sternbergorum*.[2] This honoured the Sternbergs and the fact they had been correct in their original assessment that this was a new species closely aligned to *Styracosaurus*.

'It's a really cool feeling when you find something that hasn't been described before in a museum collection – and that happens with some frequency, you know; museums are big places. And that's really the purpose of a museum collection. If you're not making new discoveries in it, it's not serving its purpose', Farke says.

It's not uncommon to browse through bones in a collection and find things that don't make sense and don't look quite like anything you've seen before, says Ryan, a palaeontologist at the Cleveland Museum of Natural History in Ohio. Furthermore 'some museums, such as the AMNH in New York and the Canadian Museum of Nature, have a deep backlog of unopened field jackets that go back more than a hundred years', he adds. 'It's all time and money to prepare these things.'

In fact, in the last few decades, many new dinosaurs – *Vagaceratops*, *Xenoceratops*, *Nyasasaurus* and *Xenoposeidon* – have all been described from specimens that sat in museum collections for many years before their significance was realised. *Nyasasaurus*, for example, dates from around 243 million years ago in the Triassic and is thought to be either the earliest known dinosaur or a close relative of it. It just goes to show, you don't always need to travel to exotic locations to make important discoveries.

An explosion of frilling finds

Several features define the horned dinosaurs or ceratopsians. These include the rostral bone above the mouth, which is part of the beak; paired horns above the eyes and a horn above the nose; a bony frill, which is an elongation of the skull that sweeps back over the neck; and frill epiossifications – small bones that form bumps or ridges around the edge.

The standard ceratopsian skull is triangular, but it has a range of gadgets added to it. These include a neck frill on all but the earliest ceratopsians, which sweeps back from the eyes. 'This varies in size. In the early forms it's pretty small, but in the later, more specialised forms it can turn into this giant sheet of bone 4 feet long, and 3 or 4 feet wide', says Farke. The other standard add-ons are a set of horns, perhaps just one above the nose in early species, and also a set above the eyes. Other species have cheek horns – spikes sticking off the side of the face formed from the bones of the cheek.

That's the basic set-up, says Farke, but some species 'got really wild and crazy. They had a whole bunch of other spikes, hooks, plates and knobs on that big bony frill. You have examples where it looks like a bunch of bananas have been cut in half and glued to the back of the frill. In others it looks like they took a pair of spears and stuck them out the back of the frill. Others have a whole bunch of little studs or knobs or curving, scimitar-shaped frill accessories'.

Since the discovery of *Triceratops*, more than 80 species have been described, something like 55 of those since 2002, indicative of a recent burst of discovery and research. 'The last 15 years we've really increased the diversity of this group', Brown says. 'In fact there are more species of horned dinosaur described in the last 10 years than the preceding 118 years. So it's a really exciting time to be working on this group.'

Peter Dodson, a palaeontologist at the University of Pennsylvania who has worked on horned dinosaurs longer than anyone else living, says that to see how dramatic the trend is, you need only consider the fact that between 1950 and 1986 no new genera were described at all. 'It once seemed a tired and dated field of study with few dedicated students. All of the ceratopsids seemed to have been discovered', Dodson wrote in *New Perspectives on Horned Dinosaurs*. 'Who could foresee the explosion of discoveries and renewed interest of the past decade and a half? There is so much material from China, Mongolia, and North America ... I see nothing but opportunity for years to come, spiced by new finds at every taxonomic level.'[3]

We're building a big database of information and starting to fill in gaps in both our understanding of the anatomy of these animals and also the gaps in geologic time between species, says Ryan. He adds that some of the things declared as new species are probably 'chrono-species', meaning they are the same animals evolving in body shape over time.

These are strange and interesting animals too, and with every new fossil there is a novel arrangement and style of headgear. Ceratopsians had the largest heads and skulls of all known land animals, both in absolute terms and relative to their body size. This was despite the fact that, for the most part, the creatures themselves did not exceed the size of a large SUV or an elephant. They were 1–9 metres long and weighed no more than 5.5 tonnes. Titanosaur sauropods may have had bodies 38 metres long and weighed 70 tonnes, but their heads were minute in comparison, with brains the size of tennis balls.

The bodies of most advanced ceratopsians (known as ceratopsids) are similar: they are all wide, thickset quadrupeds, a bit like enormous rhinos. But the head ornamentation is where evolution began to play with some beautiful creative flourishes, and the shapes and styles vary widely in each species. Some types (the chasmosaurines, such as *Triceratops* and *Chasmosaurus*) have big eye horns, small nasal horns and a large frill; while others (the centrosaurines, such as *Styracosaurus* and *Pachyrhinosaurus*) have small eye horns, large nasal horns and a shorter frill.

The great variety of ornamentation revealed in all the new species has suggested that the structures were used more for display than defence, as had traditionally been thought. Why have such a variety of different shapes of weapon when a generalised shape is presumably just as effective in incapacitating a predator?

A number of living animals have comparable structures, including chameleons and some other lizards, as well as horned mammals, such as goats, sheep, deer and cattle. 'But there's nothing that really has a combination of giant frill, with giant hooks coming off of it; and these big horns that come off the face; and these cheek horns too', says Farke. 'The thing I like about ceratopsians is that they kind of do everything with their skull. It's not just maybe a frill and maybe a horn – it's everything all at once.'

Some of the creatures with the most flamboyant ornamentation include *Kosmoceratops*, named in 2010, which Farke describes as having 'an almost Elvis Presley-like hairdo combed back over the frill'.[4] *Styracosaurus* has a series of spikes ranging from just a few centimetres long to nearly a metre, sticking out from the back of the skull. *Nasutoceratops*, described in 2013, has an oversized nose, and long, curved, forward-facing brow horns of the kind more commonly seen on cattle.[5] *Wendiceratops*, another new species described in 2015, has unusual curled epiossifications all around the edge of the frill.[6]

Clash of the Titans

Because of the prominent horns and frills of ceratopsians, there has been a lot of speculation about the function of these structures. As *Triceratops* was among the first discovered, and it had such a big frill and horns and lived alongside *T. rex*, the earliest hypothesis was that these were defensive armour. Later, the idea developed that the horns were also for battling between members of the same species, perhaps when tussling for mates. Proving this really depends on finding evidence of 'sexual dimorphism' – differences in body shape between males and females – but this is exceedingly difficult to prove in fossils, and where differences in body shape have been shown, they are usually ascribed to different species or animals of the same species at different stages of development somewhere between nestlings and mature adults.

Even more recently, experts have started to think all these horns and frills are for display, for the purpose of attracting mates, intimidating rivals and perhaps even for species recognition. 'This is consistent with a lot of these display structures that we see today in modern animals – horns and crests and dewlaps and stuff – they are generally for trying to impress or communicate somehow with members of the same species', Brown explains.

The idea of something other than predator defence for the ornamentation really took off in the 1970s, with work from Dodson and James Farlow. They looked at the fossils and saw that the horns and frills are comparable to what we see today in species such as chameleons and deer, which don't use them for predator defence. To take these ideas beyond simple analogies, Farke decided to search for direct fossil evidence of battles between males of the same species. He published a study with Darren Tanke and Ewan Wolff in 2009, in which they examined two species with different head ornamentation: *Triceratops* with its two big eye horns, and *Centrosaurus* with its more prominent nose horn.[7]

'If they are fighting members of their own species and you have animals with two different types of horn, you'd expect to see different patterns of injury in the body', Farke says. 'Something like the different patterns of injury you'd expect to see on a tennis player or a football player, because they are using their bodies in different ways. We figured the same thing would apply to horned dinosaurs, and we found that, yes, it did. There were different patterns of potential injuries when we looked at the skulls of large numbers of these animals.'

This work helped cement the idea that horns were being used for battles between members of the same species. The idea that the frills were visual display structures to show who was the biggest and most dominant – or to attract mates – is backed up by research looking at how the frills grow. Work from experts including Jack Horner, from the Museum of the Rockies in Montana, has shown that large frills develop as animals head towards maturity. 'It's only when they get a little bit older that they start adding on all the hooks and stuff that come out of the frill, and so it suggests that if it's getting more complex as the animal is getting older and closer to breeding age, then there's an important visual component.'

There's no way to know for sure if the frills were brightly coloured, but display structures today are often a vivid hue in birds (which are living dinosaurs), chameleons and other lizards, such as Australia's frill-neck. This is certainly how palaeoillustrators, such as Canada's Julius Csotonyi, are now portraying these dinosaurs in their reconstructions.

Although feathery coverings have increasingly being found in theropod dinosaurs, there's no evidence at this stage that ceratopsians were covered in feathers, although the fossil record is known to throw up occasional surprises. 'It's possible, but my suspicion is – and I don't have any hard evidence for this – that they were probably just covered with skin', says Farke.

Bristle- or quill-like structures have been shown along the back and tail of *Psittacosaurus*,[8] an early bipedal ceratopsian from Asia. There's also a long-fabled *Triceratops* specimen at the Houston Museum of Nature and Science in Texas that has a very large skin impression covering a square metre or more. This shows a variety of different scale types and possibly something that may indicate tufts or places from which bristles once emerged. But since the discovery has not yet been formally published, it's very difficult to assess it, Farke says. 'All we can say at this stage is that [these scales] had some sort of funky shape. Maybe it had something coming out of it. Maybe not.'

Some experts believe the ornamentation was used by ceratopsians to recognise other members of their own species – but there are good reasons to discount this, says British palaeontologist Darren Naish, who wrote about it on his popular blog, Tetrapod Zoology, in 2013. Foremost among these reasons, is that most species manage just fine recognising compatriots without the need for flashy signposts.[9] 'One major, crippling problem with the species recognition hypothesis is that crests and other exaggerated structures have yet to be shown to play a role in species recognition in *any* of the extant animals that have them ...' Naish says. 'Species recognition does *not* appear to be an important function for the horns, frills, antlers and so on of chameleons, hornbills, cassowaries, rhinos, deer, bovids, rhino beetles and whatever other extant animals you might consider as possible analogues for ornamented dinosaurs!' Instead, these structures are used for so-called sexual selection purposes, typically to decide who gets to mate with whom. With new fossils coming in thick and fast, and many still waiting to be described, this torrent of discovery will surely provide yet more evidence to help settle this debate.

There are several other reasons why ceratopsids are unique. The first is that they were one of the last major radiations of dinosaur groups to evolve, not appearing until around 20 million years before most dinosaurs were killed off by an asteroid impact about 66 million years ago.

The second is that – once they had evolved the overall *Triceratops*-like look every child knows – fossils of these species are almost exclusively found in the western part of North America (the one exception is *Sinoceratops*, found in China in 2010). Ceratopsids evolved on a landmass known as Laramidia that sat to the west of a vast inland sea running the length of North America from the Arctic Ocean to the Gulf of Mexico. Dinosaur groups that evolved early in the Mesozoic had a much greater chance of being found all over the world, as they appeared before the supercontinents began to break up into increasingly smaller fragments. Small, early, less specialised ceratopsians, animals such as *Yinlong*, *Protoceratops* and *Psittacosaurus*, are found in China, Mongolia and other parts of Asia, and it is thought that ceratopsians evolved there before migrating into western North America. The fact ceratopsids are only found in Laramidia is merely an accident of history, as this region was fairly isolated in the later part of the Cretaceous from around 90 million years ago.

'This is good if you work on ceratopsians and you live in North America, because it means most of the specimens are right here', Brown says. 'In southern Alberta ... we are right in the heartland of horned dinosaur country.' Currently nearly 50 per cent of all known ceratopsid species are from Alberta. This means the RTMP has been involved in a huge number of the recent discoveries, and holds the important fossils of at least seven of the species discovered since 2005, as well as many specimens that are yet to be fully cleaned, prepared and described as new species.[10]

Weird in more ways than one

Fossils in Alberta typically come from Dinosaur Provincial Park in the south-east, but the skull of Hellboy was from a river in the foothills of the Rocky Mountains in the province's south-west. For that reason alone, the palaeontologists knew the specimen was important, but it wasn't until they got the block of stone back to the lab and Tanke began the 18-month-long process of removing the skull from the rock and cleaning it up (using angle grinders, chisels, hammers and air scribes, among other tools) that its bizarre anatomy was fully revealed.

Scientists often don't have any idea how weird these ceratopsids are until they prepare them, says Farke, who has been involved in a number of other discoveries. Palaeontologists want to remove enough rock to be able to transport the fossils, but not enough that they damage them. This means more careful preparation is left for the lab. 'In the field you can be left guessing what's in there', Farke adds.

Brown says that once the skull of Hellboy was prepared, it was obvious that it was a new species, and an unexpected one at that.

'Many horned-dinosaur researchers who visited the museum did a double take when they first saw it in the laboratory', he told reporters. In fact, he quipped, the specimen was so unusual that you could tell it was a new species from 100 metres away.

What they found was a skull with a halo of large and distinctive pentagonal epiossifications around the frill, 'comically small' horns above its eyes and a large nasal horn. These small horns above the eyes are another reason it ended up with the nickname Hellboy – they reminded some of the team members of a cult comic-book character of the same name, who also has horns above his eyes.

Superficially, this species looks somewhat liked a centrosaurine dinosaur, with its small eye horns and large nose horn. But a more detailed analysis of its anatomy and the age of the rocks in which it was found resulted in a surprise for the researchers. Using fossilised pollen to gauge the age of the fossil precisely and bracket the time it could have lived, Brown and his co-worker Donald Henderson came up with a date of 67.5–68.5 million years ago. This was interesting because it was shortly after the extinction of the last centrosaurine horned dinosaurs, but before the chasmosaurine relatives of *Triceratops* went extinct after the impact of the asteroid 66 million years ago.

Hellboy was certainly a chasmosaurine, related to *Triceratops*, but it had evolved to look like a centrosaurine. Brown believes it is an example of convergent evolution – where similar features evolve independently in disparate lineages of species. This was the first time this had ever been recorded in the horned dinosaurs.

In a 2015 paper in the journal *Current Biology*, which detailed their findings and described the new species, Brown and Henderson called the animal *Regaliceratops peterhewsi*.[11] This referred to its regal crown of pentagonal plates around the frill, and also honoured Peter Hews who had brought the species to the museum's attention a decade earlier. Only the skull minus the lower jaw and rostral bone of the beak was discovered – none of the skeleton. Despite not having a body to play with, the scientists were able to make comparisons with a closely related species, *Anchiceratops*, to estimate that in life *Regaliceratops* would have weighed about 1.5 tonnes and reached a length of 6–7 metres.

Peter Hews says he was greatly honoured to have been included in the species name, and hopes it will be great encouragement for other amateur fossil hunters. He returns to Oldman River every year and has been involved in the discovery of several other ceratopsian skulls near where Hellboy was found; these are currently being prepared and studied. He has also found and collected ceratopsian dinosaur tracks, which he believes are among the best examples in North America. 'I've always been interested in fossil collecting and I think dinosaurs is the ultimate in fossil hunting', he told the *Calgary Herald*.[12] 'If you can find a dinosaur bone and imagine what this used to be and the part it played in geological history, I think that's absolutely fascinating.'

More to come

While Brown and Henderson were busy preparing their paper in 2013, other RTMP palaeontologists were not resting on their laurels. In June that year there were massive floods in Alberta – their biggest in recorded history. In some parts of the west, more than 200 millimetres of rain fell in less than two days. The floods inundated Calgary, causing 26 neighbourhoods and 75 000 people to be evacuated. But the banks of many of the region's rivers were also scoured and reformed by the floodwaters, exposing untold fossil treasures. In the weeks after the floods, the museum was inundated with calls from people who said they'd found the bones of dinosaurs and other animals along the muddy riverbanks.

Starting in spring 2014, long after the waters had receded, palaeontologists Joe Sanchez and Ben Borkovic spent two seasons prospecting along 12 rivers they believed were likely to yield fossils. All up – across more than 150 kilometres of rivers covered in the first two years of a three-year project – they found an incredible 144 sites with fossils, and excavated and collected 63 groups of specimens, many of which were jacketed with plaster and returned to the museum.

These included the significant remains of several horned dinosaurs, such as the small bipedal species *Leptoceratops*, found along Oldman River; and a much bigger skull, jaws, beak and frill of an unknown species found in Callum Creek, one of the river's tributaries. Other finds included the torso and skull of a duck-billed hadrosaur in the banks of Castle River; 40-odd dinosaur footprints in sandstone along St Mary River; and hundreds of bones of mammals, fish, birds, turtles and crocodiles in a slab of rock along the Highwood River.

The majority of the fossils are thought to be from the Late Cretaceous to Early Palaeocene, around 60–75 million years ago. Eight of the specimens are currently being cleaned and studied by the researchers at the museum with a view to publication, and possibly the description of new species, and it seems there will be more to come.

While these discoveries have resulted from the discovery of new fossil beds, in other parts of the world, palaeontologists are rediscovering old dig sites that produced astonishing fossils in the past but whose exact location was unknown.

5

THE `UNUSUAL TERRIBLE HANDS´

GOBI DESERT, MONGOLIA

In 1965 a Polish expedition into the Gobi Desert found the enormous arms and hands of a carnivorous dinosaur. The size of these mysterious limbs suggested a creature on a monstrous scale – but when the truth was finally uncovered in 2014, the animal revealed was far stranger than anyone could have imagined.

We have stepped back 70 million years into the past, to the east of the northern supercontinent of Laurasia, a region that is now the Gobi Desert of Mongolia. At this time in the Late Cretaceous, the conditions are quite different from today, and the Nemegt Basin is a great seasonal floodplain, something like the Okavango Delta of Africa.[1] Shallow waterways are clogged with vegetation, and the humid air is thick with clouds of biting insects.

Wading slowly and creating a great wake behind it that startles a flock of waterbirds is a very odd dinosaur that looks something like a creature dreamed up for a *Star Wars* movie. Almost as big as *T. rex*, it is 11 metres long and weighs 6 tonnes. Its great 2.4-metre-long arms are the longest of any known two-legged animal, and they end in 30-centimetre curved claws, which it uses to gather plant

The 'unusual terrible hands'

matter and defend itself against predators. Its narrow 1-metre-long skull has a flat, toothless bill, like that of a duck, and is covered with a beak of keratin, perfect for nipping at fresh shoots. It has a deep lower jaw, housing an enormous tongue, with which it vacuums up vegetation from the billabongs and riverbeds.

Inside its stomach are thousands of rounded pebbles – which help grind up the plants it eats – and the bones and scales of small fish it has sucked up too. On its back is a structure somewhere between a sail and a camel's hump, which may have acted as an anchoring point to support its body over its powerful legs. It has broad feet and wide, flat toes that help it balance on the soft mud of the wetlands.

This is the largest of the ostrich-like ornithomimid dinosaurs, but it is quite unlike its smaller, speedy relatives such as *Gallimimus*, which have gangly legs and small heads with large eyes. In contrast, *Deinocheirus*, with its great bulk, moves ponderously, with a plodding gait. Its curious menagerie of anatomical features is completed by a long tail with a fluffy brush of feathers on the tip, which it flicks from side to side as it walks.

Unlike the majority of its siblings from the same brood of eggs, this *Deinocheirus* has not only made it to adulthood, but survived to old age at about 27 years old. In recent months it has grown increasingly tired and begun to slow – a fact that has not gone unnoticed by a *Tarbosaurus*, a large predatory dinosaur closely related to *T. rex* that stalks the same area of this prehistoric river delta. It is no match for a healthy *Deinocheirus*, with its impressive set of huge claws, but a sick or (even better yet) a dead one would provide enough meat to keep a *Tarbosaurus* belly full for weeks.

Today is the carnivore's lucky day, for the *Deinocheirus* slowly drags its great heft out of the water and onto a sandbank for the last time. Faltering, it trips and keels over, all 7 tonnes of it hitting the ground with a resounding seismic thud. Soon, attracted by the whiff of decay, the predator moves in to feed, clamping the carcass beneath its foot and tearing off chunks of flesh like a bird of prey.

Here the carcass lies for several weeks, getting picked over by scavengers, its bones bleaching in the sun, before the seasonal flood waters start to rise and bury what's left of it on the sand bar. This is where the arms and shoulder girdle, some ribs and other bits and pieces will remain through many millions of sunrises and sunsets, numerous cataclysms and a changing environment until 1965, when a Polish–Mongolian crew of palaeontologists and explorers trundles into this region of the Nemegt Basin in a convoy of Soviet-issue military trucks ...

The awesome power of the Gobi

I'm in the east of the Mongolia's Gobi Desert, a few hundred kilometres from the border with China, and several hours' drive from the massive new Oyu Tolgoi copper mine. It's not yet 3 pm, but the sun has all but vanished from the sky. Twilight has descended as wave after wave of sand is pounding against us. We gather up our trowels, brushes and other tools, and head back to our 4WD vehicles. The awesome power of a Gobi Desert sandstorm has enveloped us, disturbing our fossil-hunting activity for the second time this week.

It's April 2015 and I'm here on a week-long fossil dig expedition that is a collaboration between the Mongolian Academy of Sciences (MAS) and *Australian Geographic*, the Sydney-based magazine I work for as editor. There's a small crew of us out here, including paying volunteers from Australia and scientists from the Mongolia Palaeontology Centre, led by fossil hunter Tsogtbaatar Khishigjav. In terms of weather, April is not the best time to be out here. Sandstorms can be bad in the Gobi at any time, but April is notorious, so we are well prepared. I'm covered up with cargo pants, a beanie and a fleece with long sleeves. To protect my eyes I'm wearing both wraparound sunnies and a pair of goggles. A bandana across my nose and mouth is doing a reasonable job at keeping the dust out. The afternoon heat has been edging towards 40 degrees Celsius, so this outfit isn't comfortable, but it is necessary.

With its own harsh beauty, steeped in mysteries and legends, the Gobi Desert covers 1.3 million square kilometres of Mongolia and the Chinese province of Inner Mongolia. Home mostly to camels, gazelle, nomads – including Mongols, Uyghurs, and Kazakhs – and the odd palaeontologist, it sweeps across the entire region, from the Tibetan Plateau in the south to Siberia in the north. The sandstorms are challenging, but for the intrepid really just add to the thrill of being out in the desert for a few weeks. Led by Tsogtbaatar, we camp out in the desert for eight nights and travel to a series of sites around the eastern Gobi to prospect for and collect fossils. Mongolia has a series of significant fossil sites, some of which are becoming tourist attractions. These include the Flaming Cliffs, site of the discovery of the first dinosaur nests and eggs by AMNH expeditions in the 1920s, and Tögrögiin Shiree, where in 1971 a Polish expedition found an almost unbelievable fossil of a *Protoceratops* and *Velociraptor* locked in combat.

Mongolia is perplexing and fascinating. There are skyscrapers, luxury goods stores, fancy restaurants and flashy SUVs in abundance in the capital Ulaanbaatar, but alongside them is a huge and sprawling ger district, where 800 000 of the 1.3 million residents live in circular gers or yurts. These felt 'tents' are simple to pack up and move, and have been the traditional dwellings of Central Asian nomads for thousands of years. Move outside the capital and you quickly realise it is the only sizeable city (home to almost half of Mongolia's nearly 3 million people). In the southern part of the Gobi paved roads are few and far between, and 4WD vehicles are a necessity to travel anywhere fast. Most Mongolians continue to live a fairly traditional lifestyle, keeping camels and tending goats on the deserts and grassy steppes. The wealth seen in Ulaanbaatar has come from mining and a natural resources boom that has developed in earnest since the fall of the Soviet Union and Mongolia's own communist government in the 1990s. A good example of the boom is the Oyu Tolgoi mine, which is predicted to generate a third of Mongolia's national GDP when performing at full capacity in 2021.

At the start of our trip in the capital, we visit the fossil preparatory lab of the MAS's Mongolia Palaeontology Centre, which is stacked with fossils that expert technicians are chipping out of great chunks of sand and rock. The collection of the MAS is truly spectacular, and we get to see some of the best specimens on display (somewhat incongruously) in an Ulaanbataar shopping mall. Specimens here include all the species I've mentioned so far in this chapter, as well as oddballs such as therizinosaurs *Segnosaurus* and *Erlikosaurus*; the pachycephalosaur *Goyocephale*; armoured dinosaurs *Saichania* and *Tsagantegia*; strange alvarezsaur *Mononykus*, which has weird little stunted arms that end in a single digit; and much more besides. Of course, the specimen I am most taken with is the huge shoulder girdle, arms and claws of *Deinocheirus*, discovered in 1965. I stand in front of it mesmerised, contemplating the dinosaur to which these bones once belonged.

We uncover a lot of fossils during that week we spend in the desert, but the most exciting moment for me is when I discover the partial mandible (lower jaw) of an *Alectrosaurus*, a Late Cretaceous relative of *T. rex* that lived in the region around 80 million years ago. This 6-metre-long dinosaur was discovered in the 1920s by the same AMNH expeditions that found *Velociraptor* and the first nests of dinosaur eggs. Isolated *Alectrosaurus* teeth are common, but skeletal remains are very rare in this part of the Gobi and have not been found here since the 1970s.

A piece of unusual, pitted reddish stone was what first caught my attention. An hour of carefully brushing at the sand revealed the front left-hand side of a lower jaw, with a single dark tooth still attached. The pits along it are the places where nerves and blood vessels once connected to each of the teeth. Much of the rest of the afternoon is spent digging out a half-metre-wide pedestal of dirt around the fossil and encasing it in hessian and plaster to ensure it makes it safely back to Ulaanbaatar for more careful preparation in the MAS lab. (Eight months later, when I am sent photographs of the fully cleaned-up mandible, I can see clearly that another four teeth, yet to emerge, remain in their sockets too.)

On this trip we collect only small and important specimens. We only have a small crew of palaeontologists and students with us, and larger specimens, such as the articulated legs of a duck-billed hadrosaur, will be investigated in more detail on subsequent expeditions. Larger specimens are reburied to disguise them from poachers, and the GPS coordinates carefully scribbled down. The intense winds and sandstorms of the Gobi scour off the surface rocks each year, exposing whole new layers of fossils. There are constantly new things to find, but this also means that anything left will have disintegrated and billowed away as dust on the wind by the following year.

Aside from the abundance of fossils, the thing that surprises me most are the specimens we leave behind. At some sites we are practically tripping over pieces of hadrosaurs, but as Tsogtbaatar tells us, the MAS already has huge numbers of good specimens of these species and has no need of more for their collections. New species, unusual fossils and pieces of theropod dinosaurs are of much greater interest – particularly strange dinosaurs, which are abundant in Mongolia.

In many of the sites we visit we find the evidence of the Russian and Polish expeditions into the Gobi – bricks, wooden posts, shards of china plates, weathered leather shoes, yellowed fragments of newspaper, food cans with labels printed in East Germany, and glass jars and bottles beautifully frosted by many decades of sand abrasion – and I pause to wonder what it must have been like on those early expeditions, particularly the one that discovered the remains of *Deinocheirus*, the strangest Mongolian dinosaur of all ...

A puzzling discovery

In July 1965, a pioneering Polish–Mongolian expedition, led by Professor Zofia Kielan-Jaworowska of the Polish Academy of Science, began to work in the Nemegt Basin region of the south-west Gobi Desert. Its discovery of a strange collection of bones – a pair of 2.4-metre-long arms, huge claws, a shoulder girdle, a few ribs and vertebrae, and a handful of other pieces – was the birth of a great mystery.

A subsequent expedition report noted:

> On the 9th, Kielan-Jaworowska found at Altan Ula III complete fore limbs and shoulder girdle of enormous size (limbs 2.5 m long) of an unknown carnivorous dinosaur, belonging evidently to a new family of theropod dinosaurs. 9–11 July were spent excavating and packing the fore limbs of the carnivorous dinosaur at Altan Ula III, and packing and loading the sauropod skeleton onto the trucks. This sauropod skeleton alone took up 35 crates, many of them weighing over 1 ton.[2]

The sauropod was prepared and subsequently described as a new species, *Opisthocoelicaudia skarzynskii*. The theropod arms were also taken back to the Polish capital, Warsaw, for preparation, and in 1970 described as a very strange new species, which palaeontologist Halszka Osmólska called *Deinocheirus mirificus*, meaning 'unusual terrible hand'.

In a paper describing the fossil, Osmólska and her co-author, Ewa Roniewicz, wrote:

> Such a manus [hand], together with the length of the fore limbs, which presumably could easily reach the ground, may suggest that they could have been used in tearing dead or weakly agile prey asunder. The length and size of the fore limbs of *Deinocheirus mirificus* set it aside from the other theropods ... Apparently gigantic theropods were represented during the Cretaceous by several forms with unreduced or slightly reduced fore limbs.[3]

This referred to the fact that other well-known carnivores, such as *Tarbosaurus*, *Tyrannosaurus* and *Allosaurus*, all have very small forelimbs.

The Polish–Mongolian expeditions that led to this discovery and many others were unusual for their time in being led by women – Kielan-Jaworowska, Osmólska and Teresa Maryańska. 'The three of them have really done most of the publishing on the dinosaurs. They did remarkable things', says Phil Currie. 'There were men on the expedition but they were there for the physical aspects of it, and were not the scientific brains behind it.' Kielan-Jaworowska once told him that the three of them would 'dread the Russians coming over to their camp because they would drink so much vodka'. One way they helped steel themselves for the encounters was by drinking a little olive oil to line their stomachs.

These fossil-hunting adventures, which ran each year between 1963 and 1971, were enormously successful. A series of Russian expeditions had also started after World War II (the first of which were in 1946, 1948 and 1949). The Russians were influenced in part by the early AMNH expeditions in the 1920s, and they went to some of the same sites, but they also found new sites such as Nemegt, and they discovered fantastic specimens, including skin impressions and one very rich site called the Dragon's Tomb. There were a number of other expeditions to the Gobi around this time, including one from Bulgaria.

In terms of productivity and numbers of dinosaurs described, however, the Polish expeditions were unparalleled. During three years of fieldwork from 1963 to 1965 alone, they collected more than 30 tonnes of dinosaur and mammal fossils. The expeditions resulted in the discovery and description of many new dinosaurs – such as *Homalocephale*, *Gallimimus*, *Bagaceratops* and *Nemegtosaurus* – in addition to more than 180 specimens, along with many new species of Cretaceous mammal (mammals were Kielan-Jaworowska's area of expertise, while Maryańska and Osmólska were more interested in dinosaurs). In total, the expeditions resulted in the publication of more than 60 papers on new fossil discoveries.

Intriguingly, Kielan-Jaworowska – who passed away aged 89 in March 2015 – said that the funding for the expeditions, and the opportunity for them to take place at all, came about through a quirk of the Soviet Union, whereby nations under Soviet influence traded goods with one another and accepted each other's currencies. This meant that Poland had accrued Mongolian currency to spare and for which it had little other use.[4]

In her autobiography, Kielan-Jaworowska offered a taste of what these endeavours involved, noting the technical complexity of mounting major expeditions, which required considerable effort and a large team:

> In the case of our expeditions, which involved a dozen or so people on the Polish side and a few on the Mongolian side, we had to assemble all the requisite gear and supplies in Warsaw. The supplies required for the excavations included plaster to protect the specimens during transport, packaging, complete equipment for the camp, food for 20 people in the field for three months, a first-aid kit, trucks and small off-road vehicles, and fuel for those vehicles (at that time, there was no high-octane gasoline in Mongolia needed for our excellent Star 66

trucks, which we rented for the expedition from the Truck Manufacturing Company in Starachowice). All supplies, which filled two large freight train cars and an open railway truck for transporting gasoline, were shipped from Warsaw in March to be picked up by our Mongolian colleagues in Ulan Bator in May ... It took us two weeks in Ulan Bator to take care of all arrangements and to prepare the vehicles for the journey.[5]

The bits of rubbish and glass bottles I found are not the only evidence of the Polish and Russian expeditions in the Gobi, says Phil Bell, an Australian palaeontologist at the University of New England in New South Wales. There's a spot at the Altan Ula site they called the Café, he says. 'There are all these wonderful inscriptions carved into the rock and there are carvings of dinosaurs. They made some graffiti on the rocks and had some fun, such as drawing dinosaurs with breasts.'

There are other interesting stories about the Russian expeditions, he says. They sometimes used bulldozers to remove the surface layers of sand and soil and reach the fossils below. Unfortunately, as I saw in parts of the eastern Gobi in 2015, this often resulted in smashed-up dinosaur specimens. One story involves the remains of a small armoured ankylosaur called *Pinacosaurus* found in 1969 at a spot called Alag Teeg near the Flaming Cliffs. This is a relatively flat area, and the Soviet–Mongolian Paleontological Expedition was able to get its bulldozers in and use them to scrape up big blocks of dirt and rock containing the dinosaurs. Puzzlingly, they had found an entire herd of baby pinacosaurs, more than 30 of them, many of which were standing upright. These animals were probably mired in mud at the time they died – there are many spots in the Gobi where you can see evidence of this, and they suggest that some of Mongolia's fossils are from animals preserved almost instantly in floods or other extreme events.

The Russians took the fossils back to Moscow for study and were amazed that these perfect little pinacosaurs – each about the size of a beagle – all lacked feet, as they reported in a very brief paper on the discovery. It was only in the last decade, when fossil hunters including Currie were able to relocate the old site, that they found hundreds of isolated feet. The bulldozers had scraped off the top of the fossils and left the feet in the ground.[6]

'I found one foot and then the other foot and the hands. Then I found more next to that, and more and more', Currie says. 'Then I realised they were all these hands and feet stuck in the mud. When you look at the Russian report they say they found all these specimens the right way up ... but that they all lacked hands and feet. Probably the 30 baby ankylosaurs [we found] were the same ones the Russians had reported on originally.'

An impossible task

Currie says that in the decades following the discovery of the *Deinocheirus* arms in 1965, palaeontologists were very curious as to what kind of dinosaur they might have belonged to. 'You could look at those arms and you could tell it was a carnivorous dinosaur. But, as you know, a big specimen, like *T. rex*, has very short arms. So if you extrapolate from these 2-metre-long arms, and you start to think, if it's a carnivorous dinosaur that's related to *T. rex*, then we're looking at an absolutely fantastically sized meat-eating dinosaur.'

But this didn't seem to make sense. The fossil had similarities to ornithomimids (which, proportionally, have much longer arms than *T. rex*) and there was talk of it being one of these ostrich-like dinosaurs even in Osmólska's paper describing it, but without a better fossil it was impossible to resolve the debate.

An initial clue to the puzzle came in 2008, when Currie and his co-workers on the annual Korea–Mongolia International Dinosaur Project (KID) expeditions had the first of several fantastic pieces of good fortune. They had been searching for the original quarry excavated by Kielan-Jaworowska in 1965. They knew the Altan Ula III region of Nemegt, but not the specific location. Currie had even returned with Kielan-Jaworowska and Mongolian palaeontologist Rinchen Barsbold, who'd both been on the 1965 expedition, and they'd still had no luck. But by taking a photograph from 1965 and performing a trick Currie had previously used to relocate quarries in Canada, they were able to find the precise spot where *Deinocheirus* had been dug up.

There's been a movement in recent years to rediscover historic quarries that yielded important specimens. This is to find more geological and environmental information about these fossils, says Bell, who worked with Currie in Canada before returning to Australia in 2013. The use of photographs to relocate quarries is a practice that has been going on for some time in Alberta's Dinosaur Provincial Park, he says. 'You may have ten dinosaur specimens all found in a single area, but they are not all exactly the same age. The only way to check this is to go and find the original quarries that these animals came from.'

There are several dinosaurs for which only partial fossils were ever found or the original fossils have been lost, and the specific locations of the discoveries is unknown. *Deinocheirus* was the most enigmatic and mysterious of these, Bell says, hence the interest in finding the quarry.

'In 2008, I was with a crew from Korea and Canada. We were in the Nemegt Valley, and part of the mission was to rediscover some of these old Polish quarries', he says. 'Armed with two black-and-white photos of the original excavation and a hand-drawn map of the very broad area, we set out to try and find this quarry. It was a matter of climbing each mountain and then trying to match up the skyline in the photo with the skyline you're looking at. It's a long and slow process, and you have very little to go on.'

Even today, parts of Mongolia remain poorly mapped with few roads, meaning expeditions have to rely on dirt tracks, flat plains and 4WD vehicles. Navigational aids were much less sophisticated in the 1960s, and the Polish–Mongolian expeditions relied on dead reckoning and triangulation. This means that even with well-marked maps, sites can be hundreds of metres or even several kilometres away from where they were recorded on maps. To compound this, in the badlands of Nemegt, you need only turn around to get lost, because the landscape is so complex and convoluted. Finding a quarry that's just a few metres square is no easy thing in a big landscape – one that could potentially have also changed shape dramatically in the past 50 years.

Despite the challenges, Currie, Bell and their Korean–Mongolian colleagues succeeded in their mission. They did find some pieces, including a toe bone that had probably been on the surface at the time of the original discovery and had since been washed down a cliff. Picking through discarded dirt from the 1965 dig resulted in the discovery of shards of bone and weathered gastralia rib bones. From these fragments the scientists were able to confirm the bones were from *Deinocheirus*, and they even found bites marks on them, which Bell later established were from *Tarbosaurus*.[7] 'Clearly *Tarbosaurus* was in there feeding on the remains of this mystery dinosaur. That was a cool little addition to the story of *Deinocheirus*, which we didn't know about before', he says.

Currie has travelled to Mongolia almost every year since 1986 – the dinosaurs here are very similar to the assemblage of dinosaurs found back in his home province of Alberta; if a new type is found in either the Gobi or Alberta, the chances are a similar species once existed in the other place too. During all those years he'd always hoped he might rediscover the *Deinocheirus* quarry, so the find left him elated, even if the fossil pickings were slim. Word of the rediscovery of the quarry had got out, however, and little did Currie know it at the time, but some exciting news from Europe was coming his way.

The find of the century

It was a small pile of notes of Mongolian tughrik currency, found the following year, that finally led Currie to one of the most important dinosaur discoveries of the 21st century. In Central Asia people often leave money or food offerings at sites of religious significance or to thank the spirits when they've had good luck. This a pile of dusty red notes, stashed under a rock, allowed Currie, Yuong-Nam Lee of the Korea Institute of Geoscience and Mineral Resources, and their KID co-workers to put a date on a partly excavated quarry at a locality called Bugiin Tsav.

In the quarry they found remains of a large dinosaur they suspected was *Deinocheirus*, and the specimen was much more complete than the 1965 original, with only its head, hands and feet missing. The tughrik notes had been printed in 2002, so that told them the dinosaur must have been removed following that date. There was a chance they could find where it had gone.

Up until 2010, poaching was a major problem in the Gobi, with Mongolian dinosaur fossils selling for huge sums overseas, particularly in the United States, and sometimes ending up in the hands of unwitting celebrities, such as actor Nicolas Cage. This was despite the fact that it is illegal to collect fossils in Mongolia without a permit and none can leave the country without government permission. If poachers discover a *Tarbosaurus*, they typically remove the skull, hands and feet, as these are easy to get out of the desert undetected and eventually smuggle overseas. The rest of the specimen is often smashed up in the process of reaching these valuable pieces.

A court case in 2013, however, which saw a US fossil dealer (labelled by prosecutors as a 'one-man black market in prehistoric fossils') charged and his fossils impounded and repatriated to Mongolia, has helped turn the tide. It's now much more difficult to sell fossils from Mongolia, and Currie says palaeontologists have seen a reduction in poaching activity.

When they found this quarry in 2009 the fossil had been stripped of its flashier parts, but much of the rest of the specimen remained. Large theropods fossils in the Gobi are typically *Tarbosaurus*, but Currie's crew quickly realised this was something different. 'In the course of taking all of this smashed bone and putting it back together and wrapping all of these big blocks of vertebrae and so on in plaster and burlap, we realised that this dinosaur we had found was in fact *Deinocheirus* and that this specimen was, scientifically, probably one of the most valuable specimens we'd found in Mongolia in more than a decade', he says. 'The real aha moment was when we pieced together a busted-up arm bone and realised that it matched the same arm bones in the type specimen that the Poles collected in 1965.'

As well as revealing unknown details about the skeleton of *Deinocheirus* and confirming that the species was an ornithomimid, the fossil also offered up some other titbits about the species' biology. Modern birds use gastroliths – small, polished stones – to help grind up plant matter in their gizzards. The researchers found thousands of smooth stones in the abdominal region of the 2009 specimen, which they interpreted as deliberately ingested gastroliths.

Soon after the discovery of the new specimen, Currie and his KID colleagues realised that another partially poached fossil they had discovered in 2006 at an area of Nemegt called Altan Ula IV was also *Deinocheirus*. It was a smaller, juvenile animal, their specimen of which consisted mostly of the spinal column. Without the arms to help with identification, they had at first decided it was a therizinosaur (*Deinocheirus* was not the only weird plant eater here – it was found alongside *Therizinosaurus*, a theropod with a pot belly and very long, scythe-like claws). Ironically, Nemegt, where all three specimens were found within 60 kilometres of one another, is one of the parts of the Gobi most highly worked for fossils. 'It goes to show that, even in places like Mongolia, where dinosaurs have been collected since the 1920s, we still find new specimens all the time', Currie says.

The 'unusual terrible hands'

'*Deinocheirus* has been one of the great mysteries of dinosaur palaeontology ever since it was first discovered', adds Bell. 'People have dreamed about figuring out what this animal really looked like. So it's just an incredible stroke of luck that they found not one but two new skeletons of this animal.'

Having two specimens with different parts of the main body, meant they were able to piece together the entire skeleton of the animal using pieces missing in one to fill gaps in the other. The final crucial pieces of the puzzle were the missing head and feet, but then Currie got a phone call from Brussels.

Bringing things to a head

In 2011, palaeontologist Pascal Godefroit at the Royal Belgian Institute of Natural Sciences (RBINS) in Brussels, had been contacted by François Escuillié the director of a fossil dealership called Eldonia, based in Gannat, France. Escuillié said he had seen a very unusual specimen owned by a private collector who wanted him to prepare and mould it. Godefroit has been instrumental in relocating poached Gobi fossils overseas, and over the years people have brought him things he has been able to identify and repatriate to Mongolia.

He only had to look at the strange skull and hands of the specimen, then owned by a German collector (who Currie speculates purchased the fossil for around US$30 000), to realise it was something unknown to science. Having heard that Currie had rediscovered the original *Deinocheirus* quarry, Godefroit called him in Canada and invited him to come and have a look. Currie hopped on a plane to Belgium and was there within the week. He was 'just blown away' by the specimen, and quickly realised that not only was it *Deinocheirus*, but it was the same specimen he'd excavated in 2009. These were the missing pieces taken from the quarry by the poachers who had left the pile of Tughrik notes behind.

'What gave us a clue that it was the poached *Deinocheirus* specimen that we collected was the fact that the hand was divided between the specimen in Europe and the poached specimen we collected in Mongolia', he says. And the specimen in Brussels was missing one phalanx or toe bone, which they'd found down the cliff from the quarry. They were able to take it and fit it right into an impression where it was missing from the block of stone with the rest of the foot. Later on, when the various parts of the fossil were reunited, the neck vertebrae fitted perfectly into the space in the base of the strange skull. 'There was absolutely no question these were the same animal', Currie says.

Once they realised the importance of the fossil, Escuillié persuaded the collector to return it to Mongolia; it had apparently been through a series of hands that had seen it travel from Mongolia to China and then Japan before passing through another dealer in Europe. It was donated to the RBINS for study and then in May 2014 presented to a Mongolian government delegation including Tsogtbaatar Khishigjav and Oyungerel Tsedevdamba, Mongolia's then minister of culture, sports and tourism. At the time of writing, the globe-trotting fossil was being prepared in Korea, but the eventual plan is for the entire specimen to be displayed in Ulaanbaatar's Central Museum of Mongolian Dinosaurs, which is currently under construction.

When the full story and the new reconstruction of *Deinocheirus* were revealed in a *Nature* paper in late 2014[8] they showed a strange creature, now known from around 95 per cent of its skeleton across the three specimens. Sadly, Osmólska, who had described the species in 1970, had by then passed away, but Kielan-Jaworowska survived just long enough to see the great mystery solved.

Other palaeontologists were flabbergasted. 'A colossal, slow-moving, horse-headed, hump-backed dinosaur that looks like something out of a bad sci-fi movie', is how Stephen Brusatte, at the University of Edinburgh in the United Kingdom, described the species to *National Geographic*. 'These new specimens really solve the mystery once and for all ... and they tell us *Deinocheirus* was much weirder than anyone could have imagined.

'It looked to me like the product of a secret love affair between a hadrosaur [duck-billed dinosaur] and *Gallimimus* [ornithomimid dinosaur]', said Tom Holtz at the University of Maryland in College Park, United States.

'The study of this specimen has shown that even in dinosaurs like *Deinocheirus*, an animal that has been known for almost half a century, we can still learn surprising things about their anatomy', Currie told journalists when the discovery was published in *Nature*. 'Furthermore, it underlines the fact that even today, dinosaurs are still relatively poorly known. The fact that *Deinocheirus* is from the Nemegt Formation of Mongolia, one of the richest and most diverse dinosaur faunas known, hints that there are probably thousands of dinosaurs that we still do not know about from the majority of dinosaur localities in the world.'[9]

The find really highlights the fact that there were probably huge numbers of dinosaur species we don't know – and never will know – anything about. The Nemegt Formation in Mongolia and Dinosaur Provincial Park in Canada are two of the richest dinosaur fossil sites in the world, but there are still great gaps in our knowledge, and there would have been hundreds or thousands of different environments around the word, each with as many species as they hold today, Currie says. 'When you think about that over a 150-million-year history ... In reality we are probably talking tens of thousands of dinosaur species that have not been discovered, and those are just minimum numbers. We're just so much in the dark when it comes to dinosaurs.'

And just when we think we have a handle on how dinosaurs evolved, a new discovery comes along that turns everything on its head – again. Once we got our heads around the idea that there were dinosaurs with feathers, many experts were convinced these were largely restricted to the carnivorous theropods. And then came a startling discovery from Siberia ...

6

SCANDALOUS BEHAVIOUR AND ENFLUFFLED VEGETARIANS

SIBERIA, RUSSIA

We thought it was settled that only carnivorous dinosaurs had feathers, but now an odd little herbivore from Siberia has spread feathers right across the family tree. Almost as remarkable was the underhand tale of its discovery, which involved claims of fossil theft, skulduggery, scandal and the ignominious practice of claim-jumping.

A herd of small, fuzzy dinosaurs is moving through the foliage on the swampy edge of a lake. They chirp at one another, some occasionally shaking and puffing up their plumage. Several remain alert – peering this way and that through horsetail and seed ferns, ready to hoot in alarm if danger is spotted – while enfluffled juveniles dart back and forth between the adults' legs.

It is the Late Jurassic, about 160 million years before this region will be known as Siberia and renowned for its icy environment. Early flying relatives of birds have recently taken to the skies, and species similar to *Archaeopteryx* and *Anchiornis* glide and flit through the trees in nearby woodlands. Back on the ground, these 1-metre-long, feathery herbivores have long tails and walk on scaly hind limbs. Their eyes are large and their snouts have teeth perfect for nibbling at ferns and conifers; they have small arms that end in hands with five fingers, and are useful for grasping vegetation as they eat.

In the distance, smoke and steam languidly rise from several volcanoes. All is now quiet, but occasionally these gateways to the innards of the planet violently explode with great clouds of ash, causing mudflows or ash falls that can rapidly bury entire herds of these dinosaurs in very fine sediments.

Fossils created in this way can preserve details such as skin, feathers and fur, and the Lagerstätten or sedimentary deposits in which they are found only occur in a very few places. These include the Solnhofen limestone of Bavaria in Germany, where *Archaeopteryx* was excavated, and the Yixian Formation of Liaoning in China, the source of so many exceptional fossils of feathered dinosaurs.

That's why in 2011, when rumours of a new Lagerstätte deposit with dinosaurs in Siberia began to circulate among palaeontologists, there was quite a ripple of excitement. The significance of the discovery may go some way to explaining why palaeontologists in Moscow planned to announce the find to the world, claiming it as their own, before the team that found the fossils knew anything about it.

The dinosaurs beyond Lake Baikal

In the summer of 2010, geology professor Sofia Sinitsa of the Siberian Branch of the Russian Academy of Sciences (RAS) was conducting a survey along the grassy banks of the Olov River in the Transbaikal region of eastern Siberia, near Chernyshevsk. Several hundred kilometres north of the border between China and Mongolia, this village is situated to the east of the vast Lake Baikal, which contains about 20 per cent of the world's liquid fresh water. Sinitsa, based at the academy's Institute of Natural Resources, Ecology and Cryology (INREC) in the nearby city of Chita, wasn't searching for fossils; she was there with her team digging trenches for a geological survey. But fossils are what she found, and they piqued her interest, as dinosaur fossils are rare in Russia. Here, particularly on the wide grassy plains of Siberia, fossils are difficult to find, and it was only because they were digging trenches that they chanced upon them.

Sinitsa returned with a small team in September that year to conduct further excavations at the site and found rocks filled with bones in several of the ditches. Two Moscow-based palaeontologists Sinitsa contacted came to look in 2011, but in her view seemed unimpressed by what they said were badly preserved and fragmented remains. Sinitsa persevered regardless and continued to collect the fossils. Not being a dinosaur expert or even a palaeontologist, she realised she needed help at this point, and contacted Pascal Godefroit, an expert on early birds and ornithischian dinosaurs at the Royal Belgian Institute of Natural Sciences in Brussels. Godefroit had worked for many years on duck-billed dinosaurs and other species at several sites in Russia.

Godefroit says he was 'immediately shocked' by Sinitsa's discoveries, because dark impressions in the fragmentary remains looked very much like the simple feather-like structures seen in carnivorous theropod dinosaurs from China, which he knew had never been found in herbivorous ornithischians before. He arrived in 2012 and they began much more extensive excavations, digging and smashing their way through the sandstones and siltstones, and uncovering many more fossils as well as hints of prehistoric volcanic eruptions.

Not only did they discover dinosaur bones, but also plants, insect larvae and crustaceans, which appeared to have been fossilised in a calm freshwater environment – very probably a lake or a bend in a languidly flowing stream. The dinosaur fossils were the most spectacular and abundant of the remains, and over the three years of digging they found six skulls and hundreds of skeletons, some partially articulated. They also dug up a single tantalising tooth shed by a predatory theropod.

'Every year we are finding something unique', Sinitsa told the *Siberian Times* in 2012, when her team was still unsure as to what kind of dinosaur it had discovered.[1] 'We have more than ten preserved bits of skin. [These] remained untouched because of the volcanic ashes from eruptions 30–40 km away, which covered the skin and acted like a time capsule ... During three seasons of our work there we have gathered a very good collection of dinosaur remains. We found a three-fingered limb, and feathering. This year we got a lot of scaled tails, as well as numerous hip and shoulder bones.'

They decided to call the new species *Kulindadromeus zabaikalicus*, meaning the 'Kulinda runner from the region beyond Lake Baikal', and set about preparing a paper to be published in an international journal, which they knew their colleagues would be eager to read.[2] While the discovery of a new species is always exciting, it was the very unusual feather impressions that really had Godefroit exhilarated, because he knew it was a major find that would completely change our idea of dinosaurs. 'I was really amazed when I saw this', he later recounted. 'We knew that some of the plant-eating ornithischian dinosaurs had simple bristles, and we couldn't be sure whether these were the same kinds of structures as bird and theropod feathers. Our new find clinches it: all dinosaurs had feathers, or at least the potential to sprout feathers.'

A palaeontological scandal

The anticipation before the announcement of *Kulindadromeus* was palpable. Godefroit had been intending to give a presentation about it at the 73rd Annual Meeting of the Society of Vertebrate Paleontology (SVP) in Los Angeles in November 2013 (the world's biggest gathering of palaeontologists). An abstract of the presentation titled 'Feather-like structures and scales in a Jurassic Neornithischian dinosaur from Siberia', which had been circulating since early 2013, provocatively noted that feather-like structures had been found with bones, 'supporting the hypothesis that simple filamentous feathers, as well as compound feather-like structures comparable to those in theropods, were widespread amongst the whole dinosaur clade'.[3]

Furthermore, the abstract added, scales on the hind limbs resembled those seen on the legs of birds such as chickens today, and the authors 'hypothesised that, at the same time early feathers evolved within the whole dinosaur clade, genetic mechanisms limiting the growth of long epidermal structures on the distal portion of the hind limb and on the tail were selected'. They were arguing that the scales were a kind of failed feather, which would explain how some dinosaurs might have had both feathers and scales, and also that the early ancestors of scaly, herbivorous dinosaurs were feathery creatures.

But Godefroit had injured his leg and was unable to present the talk as planned at the SVP meeting; the assembled crowd filed out of the lecture theatre disappointed. Snippets of gossip continued to spread through the palaeontological community while it was forced to wait for an official publication, said to be coming up in a major journal. The story of the discovery of the site had been circulating since 2010, but, aside from the SVP abstract, no concrete information had yet been provided.

The further delays opened the way for claim-jumping – which in the context of palaeontology refers to the practice of describing a new species before the group that did the work to discover it has the opportunity to publish their findings. In standard scientific practice, to describe a new species, you have to officially name it in Latin and detail its distinguishing features in a publication, preferably a scientific journal. Once this has happened that name usually sticks, unless there's a good reason to discount it – such as that the species has been inadequately described or the name has previously been assigned to another species.

Six months after the SVP meeting, in May 2014, a brief paper titled 'The discovery of late Jurassic dinosaurs in Russia' appeared in the Russian journal *Doklady Earth Sciences*, briefly describing the new finds at Kulinda.[4] The author was Vladimir Alifanov, of the RAS in Moscow. That was followed by another paper by Alifanov and his colleague Sergei Saveliev titled 'Two new ornithischian dinosaurs (Hypsilophodontia, Ornithopoda) from the Late Jurassic of Russia'.[5] In it they described two species they called *Kulindapteryx* and *Daurosaurus*, and they made no mention of Godefroit or Sinitsa.

The plot thickened. Other members of the community who already knew of Godefroit's SVP abstract were confused. Had Alifanov found a different species at Kulinda? Was he working on a different site? Or had he, as many suspected, claim-jumped Godefroit and Sinitsa to take the glory of naming the new species for himself?

An answer of sorts came when a comment published on Facebook by Godefroit was republished and circulated by email on the widely used Dinosaur Mailing List in the first week of July:

> Please forget about those names. These are based on specimens that Alifanov stole in Chita and that are illegally housed in PIN [Palaeontological Institute] in Moscow but belonging in fact to the Institute of Natural Resources Ecology and Cryology (Chita). Moreover there is a single taxon. And it is not a hypsilophont but a basal ornithischian.

> This paper is a true paleontological scandal! The official paper implying the original discoverers of the specimens ... and based on the legal INREC material in Chita will be published in July in one of the highest-ranked scientific journals![6]

So it seemed that Alifanov and Saveliev had set out to describe the species ahead of the authors of the *Science* paper, which wouldn't appear in print until 25 July.[7]

Mike Benton at the University of Bristol in the United Kingdom, another co-author of that paper, confirmed that it was Sinitsa, her colleague Yuri Bolotsky and other Russians from Novosibirsk that had first found the site at Kulinda and managed the excavations there. It was Alifanov and Saveliev who had come to have a look at the site back in 2011, he says, spending a 'fraternal day or two' there but seeming unimpressed with the fossils and swiftly returning to Moscow. According to Benton, the team discovered afterwards that some fossils had been taken without permission, 'a real no-no in any country or situation. Maybe they had discussed having a few samples for the PIN collection – but in fact it was really a theft. I think the Russian excavation leaders feared something was up because Alifanov had been a little odd and had said very little'.

In their paper, Alifanov and Saveliev wrote that they had been on a 'joint expedition' with INREC, but they made no mention of Sinitsa whatsoever, even though she was the project leader who found the site and was managing excavations there. Looking on from outside of Russia, this behaviour seems perplexing, but Benton says it may simply reflect what was previously normal working practice in the era of the USSR, when everything was centralised around Moscow. For many years, it was normal for scientists at the Palaeontological Institute of the RAS to come out to regional centres and request that collections of important fossils be sent to Moscow, and local scientists accepted this largely without complaint.

'Alifanov's paper appeared in Russian a few months before our *Science* paper', Benton says, 'and then in an English translation edition also before ours. Formally, they named two dinosaur genera, each based on rather inadequate materials, so the two names become instant nomina nuda. This is a huge pity, and quite unnecessary – if only everyone had decided to work together, under the leadership of Sinitsa and Godefroit'.

Nomina nuda (singular 'nomen nudem') means 'naked names' and is used to refer to Latin species names that have failed to stick because the species were inadequately described in the initial publication, with too little information to distinguish them from other named species. In their paper, the RAS scientists, with their limited collection of specimens, claimed the fossils represented remains from several species, while Godefroit's team said there was only one. A subsequent paper, published by Alifanov and Saveliev in 2015,[8] claimed a third species, an ovimimosaur, from the site, but other palaeontologists pointed out that this fossil of a supposed hand bore a remarkable similarity to the foot of the *Kulindadromeus* specimens still housed at INREC in Chita.

So now there are three possible names for the fuzzy Kulinda ornithischian floating around, and it may be that one of Alifanov's names comes to stick. It seems, however, that most palaeontologists and dinophiles have already come to a consensus and are calling the species *Kulindadromeus*.

Creeping enflufflement

To understand why the discovery was so significant, you have to know a little bit about the way dinosaurs are categorised into different groups. The dinosaur family tree is split into two major branches: the saurischians, which include all the carnivorous theropods and the long-necked sauropods; and the ornithischians, which includes all the armoured dinosaurs, the horned ceratopsians and the herbivorous duck-bills.

Beginning with the shock 1996 discovery of a fuzzy little species called *Sinosauropteryx*, we now know of about 50 species of feathered dinosaur, the majority being theropods from the north-eastern Chinese province of Liaoning, about 1500 kilometres south-east of Kulinda. These range from the magpie-sized gliders *Microraptor* and *Anchiornis* to the 10-metre-long shaggy *Yutyrannus*.

Before the discovery of *Kulindadromeus*, just two ornithischian dinosaurs – *Psittacosaurus* in 2002 and *Tianyulong* in 2009 – had previously been found with bristle- or filament-like structures potentially related to feathers. The lack of feathers in these herbivores, and fossilised scaly skin impressions for some species, particularly hadrosaurs, was why many believed feathers were restricted to theropods and not widespread across all dinosaurs.

Most experts accepted that the structures found in *Psittacosaurus* and *Tianyulong* were made of keratin, (as are scales, feathers, fur and quills), but the question was whether they were actually a kind of feather from a distant ancestor that shared an origin with the feathers of the theropods, or whether they had a separate evolutionary origin, meaning that feather-like structures had evolved on multiple occasions. The oldest feathered dinosaurs we know of so far are around 160 million years old, but if feathers evolved just once and pre-dated the split between ornithischian and saurischian dinosaurs, then they must have first appeared more than 200 million years ago.

The numerous well-preserved fossils of *Kulindadromeus* have brought experts much closer to answering this question. These specimens – which Sinitsa's geological work suggests were somewhere between 144 and 169 million years old – preserve a number of types of both scales and feathers, which was a first.

Kulindadromeus had neat rows of large scales along the featherless tail, hexagonal scales on the lower legs, and rounded scales around the forelimbs and ankles. To go along with the three types of scale, there were three types of feather: tufts of fluffy filaments on the tops of the arms and legs, more simple single-filament bristles around the head and body, and odd ribbon-like feathers on the lower legs that have never been seen in any other creature, extant or extinct. 'These feathers are really very well preserved', said team member Maria McNamara now at University College Cork in Ireland. 'We can see each filament [on the legs] and how they are joined together at the base, making a compound structure of six or seven filaments, each up to 15 millimetres long.'[9]

Experts who were not involved in the discovery seemed split on the significance of the find. Stephen Brusatte, a palaeontologist at the University of Edinburgh in the United Kingdom, told *Science*: '*Kulindadromeus* seals the deal that some plant-eating dinosaurs had feathers, and is the best proof yet that feathers weren't something that evolved only in the meat-eating dinosaurs ... It tells us that feathers must have arisen earlier in dinosaur evolution than most of us previously thought, and maybe even the common ancestor of all dinosaurs had feathers'.[10]

Paul Barrett of the Natural History Museum in London wasn't so sure the marks were feathers. 'Most feathers have a branching structure', he told the BBC. 'Instead these look like little streamers coming from a central plate. No bird has that structure in any part of its plumage and none of the developmental models that biologists use to understand the evolution of feathers includes a stage that has anything like that kind of anatomy.'[11]

After the initial discovery and analysis of the fossils, Benton and McNamara, then also at the University of Bristol, were brought on board to study them; both had previously worked on probing fossils for tiny packages of pigment called melanosomes. These account for the colour of feathers and fur, and have been found fossilised in many Chinese specimens of feathered dinosaurs and birds. In 2016, Godefroit confirmed that melanosomes had been found in the *Kulindadromeus* fossils.

'The biggest question still is about the origin of feathers and feather-like structures. We want to be sure the structures we can see in *Kulindadromeus* are really related to the development of feathers in dinosaurs. That's why we are trying several different kinds of examinations to prove those structures are really related to feather development in dinosaurs', he says. 'We have studied the feathers of *Kulindadromeus* and we could identify melanosomes as seen in *Microraptor* and [other theropods], so these are really protofeathers, they are epidermal appendages.'

Though they now have evidence of melanosomes, they are not sure if their fossilised shape corresponds well to the shape of these packages of pigment when the dinosaurs were alive. 'That's the problem for identifying colour. Maybe we will be able, but for the moment it's really too early', Godefroit says. The team is currently working on the melanosomes to see if they can reveal the colour of and also any patterns in the plumage.

The presence of both feathers and scales in the fossils shows that these two kinds of skin-based appendage could exist together in dinosaurs, as they do today in chickens with scaly legs and vultures with bald, scaly heads. Developmental experiments in chickens performed by Danielle Dhouailly at the Université Joseph Fourier in Grenoble, France, have suggested that the leg scales are in fact a form of aborted feather that has halted development. 'The astonishing discovery is that the molecular mechanisms needed for this switch might have been so clearly related to the appearance of the first feathers in the earliest dinosaurs', says Dhouailly, another of the co-authors of the *Science* paper.[12]

The scientists believed *Kulindadromeus* showed that the ability to produce feathers may have evolved early in the history of dinosaurs, perhaps in the Triassic, more than 230 million years ago. These feathers were not used for flight until much later, in the theropod ancestors of birds, and most dinosaurs used them for insulation and display. Most herbivorous dinosaurs may not have had feathers, as suggested by the skin impressions in some fossils, but these feathers may have been lost as animals grew to large sizes, much as the largest land mammals today – including elephants, hippos and rhinos – have lost the majority of their covering of fur. Whether they are called feathers or bristles in a way does not matter. These are clearly not feathers in the style of modern birds, but it is likely that genetically they share an origin with both the feathers of birds and the hair of mammals.

The fossils from Kulinda are significant for several other reasons. Non-marine fossil sites from the middle Jurassic are rare, having previously only been found in China and England. The discovery of a range of fossils from an ecosystem including dinosaurs, plants and insects is a remarkable boon to palaeontologists, and it will take years of detailed study before they reveal all their secrets. Furthermore, because the dinosaur specimens come from hundreds of animals and cover many different ages and stages of development, they offer an unparalleled opportunity to look at how feathering and other features changed as the animals grew and developed.

Understanding how these dinosaurs were fossilised is another question the scientists are hoping to answer. 'This is a very unique fossil site in that the bones are not well preserved, but the integumentary [skin-based] features are, and for the moment we really don't understand what happened here. It's a mystery', Godefroit says. To solve this, he has a PhD student working on uncovering the geology and sedimentology, and what happened to the bones after the animals died.

The team particularly wants to understand the strange mixture of sediments and volcanic ash in these rocks, and also the common features in the fossilisation process here and in the Yixian Formation of Liaoning in China, famous for its incredibly detailed preservation (see chapter 2). 'In the Yixian, there is the same mixture of volcanic ash and sediments, but nobody has really investigated the mineralogy and sedimentology', Godefroit says.

Revealing Mesozoic Siberia

Though it was the most exciting recent discovery, *Kulindadromeus* is not the only evidence of dinosaurs found in Siberia of late. Kundur in the Amur Oblast, to the south-east of Kulinda, has yielded a series of specimens of dinosaurs including the duck-billed hadrosaurs *Kundurosaurus nagornyi* and *Olorotitan arharensis*, described by Godefroit in 2003 and 2012. *Orolotitan* means 'gigantic swan', and the species is distinguished by an unusual head crest that sweeps back from the skull in a fan-like shape. These dinosaurs, found near the border with China, date from around 66–70 million years ago, at the very end of the Cretaceous.

'We found very big bones beds with thousands of hadrosaur skeletons – across four sites in China and Russia we identified six or seven genera of hadrosaur', says Godefroit. 'It was very interesting, because these are the same age as the Hell Creek *T. rex*–bearing localities in America, but the composition of the fauna is completely different.' At Hell Creek, palaeontologists are tripping over the bones of ceratopsian horned dinosaurs, such as *Triceratops*, but here in the Amur region of Russia, which in the Cretaceous was joined to North America, they didn't find a single horned dinosaur bone.

Similarly, over the last two decades, palaeontologists at Tomsk State University in Russia have been regularly finding fossils along the Kiya River in south-western Siberia. There in the Kemerovo Oblast, fossils regularly emerge from steep riverbanks near the village of Shestakovo, about 1500 kilometres west of Lake Baikal.[13] They have revealed a diverse assemblage of birds, mammals, fish, amphibians, turtles and crocodiles, as well as dinosaurs such as dromaeosaurs and troodontids. Removing bones from 10 metres up in the coloured bands of hard sandstone can be challenging and sometimes requires explosives. Sadly, in Shestakovo, as in some other regions in Russia, there has been a problem with fossil poaching, with specimens bought from locals and sold to overseas collectors on the black market. The very small number of scientists working on dinosaurs in Russia, and the lack of funding for palaeontology has exacerbated the problem.

Among the fossils found in 2008 were large bones dug out in huge sandstone blocks and taken to the Tomsk University's palaeontological museum, where preparators spent seven years cleaning and revealing them. In March 2015, a team led by Stepan Ivantsov from the Laboratory of Mesozoic and Cenozoic Continental Ecosystems reported that they had found a new titanosaur – the first time this type of sauropod had been seen in Russia. The bones, which included a shoulder blade, some vertebrae and a sacrum, are among the best sauropod remains from Siberia, which experts had previously branded a 'terra incognita' for sauropod evolution.

'When we discovered this find, it was only clear that the remains belonged to a very large herbivorous dinosaur from the sauropod group ... the first scientifically described dinosaur from this [titanosaur] group in Russia', Ivantsov told the *Siberian Times*. 'Now after work on the extraction of all the remnants and the restoration are almost completed, we can confidently say that we have found a new species, and maybe even genus.' It's likely that a foot found in 1995 is from the same species. All the bones date from around 100 million years ago during the Mid-Cretaceous.[14]

The dinosaur graveyard at Shestakovo has been the site of a series of discoveries of dinosaur bones and teeth, purported to include a male and female pair of parrot-beaked psittacosaurs (small and early relatives of *Triceratops*) killed together by a falling tree. More recently, Natalya Demidenko of the nearby Kemerovo Museum led excavations of several adult and juvenile specimens of *Psittacosaurus sibiricus*.[15] Announcing that find in 2014, she told the *Siberian Times*: 'We found not only one skeleton which in size is 2 metres in length and 1.5 metres wide, but nearby saw one smaller skeleton and two little skeletons of cub dinosaurs'.[16] This species of *Psittacosaurus* was perhaps twice as large as Mongolia's *Psittacosaurus gobiensis*, growing up to 3 metres long.

Godefroit has been involved for a number of years in work in another area of Siberia – the Kakanaut River in the Koryakia region of the Kamchatka Peninsula, which juts down between the Sea of Okhotsk and the Bering Sea. From the eastern coast of Kamchatka, the next landmass you meet sailing east is the US state of Alaska. The fossils at Kakanaut were first discovered in the Soviet era by a pioneering husband-and-wife team Lev Nessov and Lina Golovneva. From their base in St Petersburg (then Leningrad), and with very little money or logistical support, this vertebrate palaeontologist and palaeobotanist pair travelled to remote regions of the USSR, often hitching rides with locals or travelling by train. In places such as Kamchatka and Kazakhstan they discovered new fossil-bearing localities and a host of new species.

As Roland A Gangloff details in his book *Dinosaurs Under the Aurora*: 'They and their assistants exhibited a tenacity and dedication to their science that defies adequate description ... In 1989, Lev and Lina found scattered dinosaur bones in part of a thick sequence of continental and marine volcano-sedimentary rocks that crop out along the Kakanaut River.'[17]

The dinosaur remains found here over the years are mostly teeth, but also include some bones, and they reveal the presence of species such as duck-billed hadrosaurs, horned dinosaurs, tyrannosaurs, dromaeosaurs, ornithopods and armoured ankylosaurs. All of these remains are from the very end of the Cretaceous, 66–68 million years ago, just before the mass extinction event that saw the extinction of the non-bird dinosaurs. 'The material was fragmentary but showed that the Arctic dinosaurs were very diverse', Godefroit says.

The Kakanaut finds have been important in understanding the end-Cretaceous extinction event. This is in part because Kakanaut was within the Arctic Circle and just 1600 kilometres from the North Pole at that time. Conditions then were warmer than today, but mean annual temperatures were still around 10 degrees Celsius, and there would have been frequent spells below freezing and many months of darkness. Dinosaurs fossils found across the Bering Sea in Alaska were left by creatures that endured similar climatic conditions, but some experts have argued that they migrated south in winter, en masse, to avoid the coldest months. At Kakanaut, Godefroit, Golovneva and their co-workers found fragments of eggshell from hadrosaurs and theropods, which suggested that these animals were breeding in polar regions and living there year round.[18]

Some scientists have claimed that falling global temperatures had led to a decline in dinosaurs around the world in the run-up to the impact of the massive asteroid or comet that created the Chicxulub Crater on Mexico's Yucatan Peninsula. But the rich fauna found by Godefroit and his colleagues suggests that not only had a diverse ecosystem of dinosaurs persisted in the Arctic in the Late Cretaceous, but also that they were thriving in very cold conditions. The herbivores here must have fed on evergreens such as conifers and also taken advantage during summer months of a profusion of nutritious fresh growth thrown out by plants bathing in light 24 hours a day.

'For the first time we have firm evidence that these polar dinosaurs were able to reproduce and live in those relatively cold regions. There is no way of knowing for sure, but dinosaurs were probably warm blooded just like modern birds, which are the direct descendants of dinosaurs', Godefroit told reporters when their findings were published in the German journal *Naturwissenschaften* in 2009. 'The dinosaurs were incredibly diverse in polar regions – as diverse as they were in tropical regions. It was a big surprise for us.'[19]

Rather than dinosaurs slowly dying out due to climate change in the run-up to the impact, Godefroit believes that the discovery backs up the idea that dinosaurs were killed off in a rapid and brutal fashion by cataclysmic conditions that swept the world following the Chicxulub impact. Debris in the atmosphere may have blackened the skies for several years, killing off plants and destroying the food chain from the bottom up – particularly as large herbivores, such as sauropods, required vast quantities of plant matter to fuel their massive bulk. Starved of meat, the flesh eaters would eventually have succumbed too.

It's likely that a combination of factors led to the demise of the non-bird dinosaurs, but the precise explanation remains a fascinating and enduring mystery. It was long thought that Siberia had a paucity of fossils in comparison to its southern neighbours Mongolia and China, but the recent discoveries of species such as *Kulindadromeus* suggest this may not be true. Sadly, since the demise of communism and the USSR, Russia has very few active dinosaur palaeontologists. There is hope that the high-profile new discoveries will inspire a younger generation to join their ranks and reveal a fresh trove of untold treasures to the world.

Back in Kulinda, Godefroit and Sinitsa continued to excavate in 2015, and had by then found *Kulindadromeus* in three bone beds, but – aside from the single theropod tooth – they now believe this is the only dinosaur species to be found there. The goal now, Godefroit says, is to search the surrounding area for new sites of the same age that may yield more spectacularly preserved specimens of other dinosaur species.

The rate of discovery of new dinosaur fossils in has increased in Siberia in recent years, but in similarly chilly conditions on the opposite side of the Bering Sea in Alsaska, American palaeontologists have found an even more impressive trove of dinosaurs from the far north.

7

CRETACEOUS CREATURES OF THE FROZEN NORTH

ALASKA, UNITED STATES

We used to think of dinosaurs as cold-blooded reptiles that couldn't survive the winter cold. But thousands of dinosaur footprints at numerous sites across Alaska, and the fossils of new species from the state's North Slope, show us that a rich diversity of polar dinosaurs was not simply subsisting, but was booming across the Arctic during the Late Cretaceous.

It was the summer of 2007, on the last day of an exploratory expedition for traces of dinosaurs in the Alaska Range of mountains spanning Denali National Park. Anthony Fiorillo of the Perot Museum of Nature and Science in Dallas, Texas, was waiting for a rubber mould of a bird footprint to set. Meanwhile, his colleagues, Stephen Hasiotis of the University of Kansas and Yoshitsugu Kobayashi of the Hokkaido Museum in Japan, went exploring further up the rocks of the Cantwell Formation. Before long, Yoshi returned breathless and almost in tears, crying, 'Tony, you have to drop what you're doing and come see this'. Fiorillo, doing his best to exhibit the level-headed discipline of a scientist, refused and lingered for another 20 minutes, Yoshi trying all the while to drag him away.

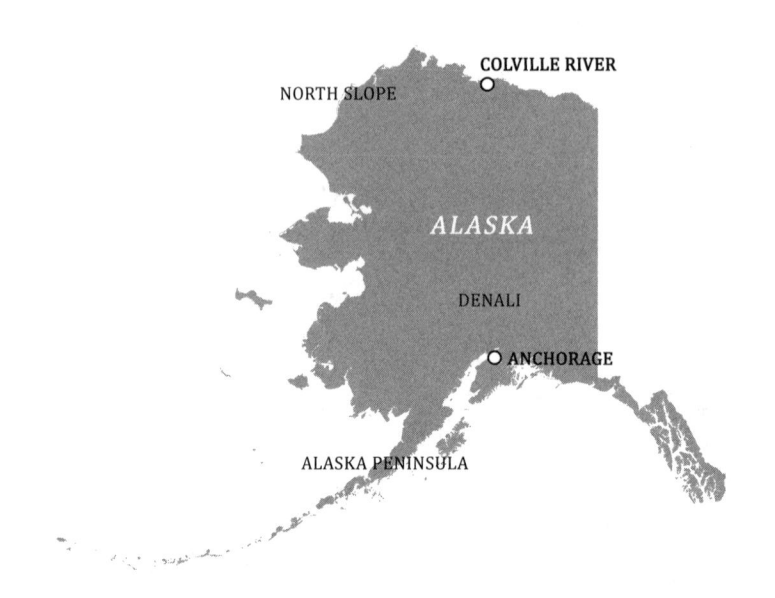

When he finally did round the bend, his mouth fell open and he felt giddy with joy and excitement. The track site was enormous, big enough to fill a football field, with thousands upon thousands of dinosaur footprints. He had never seen anything like it before. Some of the tracks were so perfect that they included the impressions of scaly skin from the feet. Some were left by running or walking dinosaurs, others revealed where they had slipped in the mud.

Lying in the remote interior of Alaska, the national park is named for North America's highest peak, Denali, or Mt McKinley, which rises to 6190 metres. The park's 2.4 million hectares of taiga forest, alpine tundra and snow-smothered summits has just a single road, and an average temperature of minus 2 degrees Celsius, making this a stunning but very challenging place for palaeontology. Clearly, though, persistence pays off.

The trio published a paper on their find in the journal *Geology* in 2014.[1] The footprints, two-thirds of which are from duck-billed hadrosaurs, were left about 70 million years ago in the Late Cretaceous. These footprints range in length from 8 to 64 centimetres, and the scientists interpreted them as being left by a herd of duckbills of different ages, from babies and juveniles to adolescents and full adults. The fact that there were very small numbers of the tiniest tracks suggested that the baby dinosaurs grew very rapidly and also – most significantly – that these were family units of a species that lived here year round. It had previously been assumed that polar dinosaurs migrated to warmer climes in winter and back up to the Arctic in summer, but these animals were clearly permanent residents.

The tracksite of hadrosaur families is just one of many streams of evidence suggesting that dinosaurs were abundant and successful in the Arctic during the Cretaceous period. And the finds are significant, because they paint a very different picture from the now dated idea that dinosaurs were cold-blooded, slow-moving creatures similar to modern reptiles. These polar animals with high metabolic rates, some with insulating coverings of feathers, were keenly adapted to their Arctic environment. They include the hadrosaur *Ugrunaaluk*, described in 2015, and the pygmy tyrannosaur *Nanuqsaurus*, revealed in 2014.

'The more we look, we are finding that dinosaurs were really widespread in the Arctic. Not just through one or two windows in time, but through long periods when Alaska was actually farther north than today', says Patrick Druckenmiller, a palaeontologist at the University of Alaska Museum of the North in Fairbanks. 'Ten years ago, most people didn't even think dinosaurs were living in the Arctic. Now we're finding that dinosaurs weren't just surviving in the Arctic – they were thriving in the Arctic.'

Seventy million years ago, Alaska was warmer than today, with an average temperature of perhaps 6 degrees Celsius and a climate comparable to the modern-day US Pacific north-west. The big limitation for plants and animals then wasn't the cold, but the 3–4 months of winter darkness. There were vast polar forests here then, an ecosystem with no modern equivalent. During the winter months, plants and photosynthesis would shut down, broad-leafed trees would lose their leaves and conifers would drop their needles. For animals here in summer, the living was great, as trees and shrubs exploded with verdant foliage in the 24 hours of sunlight, but winter would have been a lean time.

'One of the things we'd like to understand is how these dinosaurs survived. Did they migrate? Did they stay put? Did some of them even hibernate?' says Druckenmiller. 'Most of us agree they were probably not migrating. The distance these animals would have to migrate just to reach the Arctic Circle at that time would have been much farther than any land mammal migrates today, so it doesn't seem likely they were travelling that far. They must have had some way to survive on low-quality forage in the long winter months.'

Plant eaters perhaps survived partly on what they had eaten in the summer and partly on vegetation of low nutritional value: horsetails and other ferns, bark, branches and rotting wood. Moose do this today in the Arctic, and also, because they are relatively large animals, survive on fat reserves from the good times. Other large mammals, such as humpback whales, are able to fast for many months, surviving on what they ate in summer. Perhaps the herbivorous dinosaurs could do this or drop their metabolic rates and food requirements in winter.

'We don't know if dinosaurs could hibernate, but there's no particular reason why they shouldn't have been able to, as some modern reptiles and mammals do', Druckenmiller adds. But much of this remains guesswork, and the details of how dinosaurs survived at high latitudes in cold conditions and with months of darkness remain to be seen.

A palaeontological candy store

'The first dinosaur footprint at Denali was found in 2005, but now we have thousands and thousands of footprints, and they are gorgeous and incredibly diverse', Fiorillo says. 'And they are almost exactly the same age as the bones on the North Slope ... the footprints at Denali tell us what the dance was, and the bones on the North Slope tell us who was dancing.'

Since the first fossil-hunting expeditions to the North Slope region in the far north of Alaska in the 1980s, tens of thousands of dinosaur bones have been found in addition to all the footprints. Dinosaur palaeontology had a slow start in Alaska, but 'we went from zero to 60 in a very short window of time', Fiorillo says. The North Slope is the best place for dinosaur bones, while a series of sites elsewhere in the state offer numerous dinosaur footprints.

The first print found at Denali in 2005, near a site called Igloo Creek, was left by the three-toed foot of a carnivorous theropod. Five different types of theropod have now been recognised in the footprints, including a really weird one that was a first for North America.

A four-toed print found in 2012 on a remote mountain slope is the unmistakable impression of a therizinosaur dinosaur.[2] Therizinosaurs, such as *Beipiaosaurus* and *Therizinosaurus*, mostly known from China, are strange theropods that evolved into herbivores with very long scythe-like claws. These relatives of *T. rex* were similarly huge dinosaurs that walked on their hind limbs and were covered in a shaggy coat of feathers. They had slender, forward-facing toes on their hind feet, peg-like teeth for stripping branches of vegetation, wide bodies with pot bellies, and what Fiorillo describes as knuckle-draggingly long claws.

The first therizinosaur known from North America, *Nothronychus*, was found in New Mexico in 2001 (a second, *Falcarius*, was described in 2001), but the Denali discoveries – which now include a number of trackways – were the first outside Asia. 'The footprints match up very clearly with the anatomy of the feet of therizinosaurs, so there's little doubt in my mind that that's what they are', says Fiorillo. He is hopeful they will find the fossilised bones too.

And really, this is just scratching the surface of what was living in this polar ecosystem.[3] Fiorillo has also found hand- and footprints of pterosaurs with wingspans ranging from 6 to 12 metres, animals that were similar to *Pteranodon* or the biplane-sized *Quetzalcoatlus*. Apart from the tracks left at Denali by hadrosaurs, there are plenty from horned dinosaurs too, possibly left by an 8-metre-long, 4-tonne species discovered by Fiorillo in a North Slope bone bed and named *Pachyrhinosaurus perotorum* in 2012.[4] The footprints were left in a variety of environments, including lakes, rivers and floodplains, and even uplands.

They have also discovered the footprints of an unprecedented number of Cretaceous birds, which lived alongside the dinosaurs and likely endured epic migrations to reach their summer breeding grounds here each year. Fiorillo has tracks from seven different birds, ranging in size from a whooping crane to a shorebird. Other tracks from invertebrates are informative about the climate and seasons. 'Denali may not have bones, but the track record is so incredible here that it fills in a lot of the missing information from the big bones beds of the North Slope', he says. 'It is a palaeontological candy store.'

Very few places have such an abundance of dinosaur tracks alongside sites with abundant dinosaur bones of around the same age. There are remains of dinosaurs in Antarctica and parts of Australia that were at far southerly latitudes during the Cretaceous, but there is no doubt that Alaska is currently the best place to study the mysteries of polar dinosaurs.

The mighty Yukon

In July 2013, Druckenmiller had his own magical experience with another incredible new track site in a previously unexplored part of Alaska. Poring over geological maps, he had seen many regions with rocks of the right age that he wanted to explore. It was clear from the maps that the Tanana and Yukon rivers, in the central west of Alaska, would slice right through dinosaur-era sandstones and mudstones. 'We started to ask ourselves if there might be dinosaur tracks on the Yukon', Druckenmiller says. Along with a team from the Museum of the North, he embarked on an incredible 800-kilometre journey by boat down these Tanana and Yukon rivers, collecting 900 kilograms of fossilised dinosaur footprints along the way. Getting there and mounting an expedition on these remote rivers was an enormous logistical challenge, but it was very much worth the effort.

'It's was really cool after all this planning to hop out of the boat on a beach with rocks that looked about right ... You start to flip over rocks and think, "Hey, wait, that might be a dinosaur track, and that might be a dinosaur track", and eventually you find one that makes it painfully obvious that without a doubt that's a dinosaur track.' As they worked their way along the Yukon, they slowly began to realise what they had found. The footprints were so abundant at some of the sites that they could count as many as 50 of them in just 10 minutes of searching. It was a huge and important new site for dinosaurs.

'It's not always like you crack open a rock and bingo, there's the thing. Sometimes it's a dawning realisation that sort of explodes in your brain over a couple of hours or a day', Druckenmiller says. 'You realise that you're the first person to recognise that an area that was not on the map for dinosaurs the day before is now an incredibly important field site.'

The footprints here are natural casts created when sand and sediment filled holes left by the feet in the mud, and they may be 25–30 million years older than those found 425 kilometres to the west in Denali National Park.

Denizens of the far north

The first dinosaur bones to be found in Alaska were discovered in 1961 along the Colville River of the North Slope in the state's far north, about 50 kilometres from where the water flows out into the Beaufort Sea. Robert Liscomb, a geologist with Shell Oil, was mapping the Colville when he found a number of well-preserved hadrosaur bones. He didn't know what they were, and intended for them to be studied by a palaeontologist, but, unfortunately, was killed the next year in a rockslide. The bones were all but forgotten for the next few decades.

The bones remained in storage at Shell until they were discovered in the early 1980s and passed on to the US Geological Survey at Menlo Park in California for further study. After spending a stint mislabelled as mammoth bones, they were recognised as dinosaur bones by palaeontologist Charles Repenning, who immediately understood the significance of this Alaskan discovery. Eventually, in the late 1980s, an expedition from the University of California, Berkeley and the University of Alaska Museum of the North was able to use Liscomb's notes to sleuth out the original location of what is now called the Liscomb Bone Bed.

A part of the Prince Creek Formation of rocks, deposited on an Arctic coastal floodplain about 70 million years ago, these beds contain some of the northernmost dinosaurs ever excavated. Today this fossil site is found more than 1000 kilometres north of Alaska's most populous city, Anchorage, at about 70 degrees North – but in the Late Cretaceous it was even nearer the North Pole, at about 80 degrees North.

The bone beds, layers of rock 60–90 centimetres thick at the base of the river's cliffs, are packed full of a jumbled mixture of dinosaur bones, the majority from juvenile duck-billed hadrosaurs that were about 3 metres long. The experts are unsure how extensive the beds are, but they go at least for a few hundred metres in either direction. 'There are thousands of bones. We have 6000 of them at the University of Alaska now', says Gregory Erickson, a dinosaur palaeobiologist at Florida State University in Tallahassee. 'The animals range from young individuals to adult size, so it gives us a good snapshot of what a population of animals was like.'

The experts think that these dinosaurs were killed in some kind of calamity, rather than fossilising at different points over a longer period. 'For some reason a large number of these animals all died at one time. And it's river and stream sediments, so it seems to have something to do with that setting, as opposed to say a drying waterhole', Erickson says. It could be that they were trying to cross a flooded river, some of them got caught up in the current and drowned, and then their bodies were washed together and became fossilised. The bones are scattered and for the most part disarticulated, although partial skeletons are occasionally recovered. These are elephant-sized animals, so for them to have been moved by the river and buried, there must have been a lot of energy in the water and a lot of sediment to cover them. Fiorillo believes the dinosaurs here were killed in flash floods. High mountains close to coastal plains in an Arctic region subject to rapid thawing of snow and ice when the sun returned each year could have meant flooding events that were huge in scale and capable of sweeping whole herds of dinosaurs to their deaths.

One of the new species discovered by studying the bone beds is the hadrosaur *Ugrunaaluk kuukpikensis*, which means 'ancient grazer of the Colville River' in the Iñupiaq language of the local indigenous Alaskans. This 9-metre-long herbivore had hundreds of small teeth packed into large dental batteries that were perfect for grinding vegetation and would wear down as they fed. These herd-living animals could drop down onto all fours or stand on their hind legs, and had large, broad bills used for cropping ferns, horsetail ferns and the Cretaceous-era equivalent of duckweed.

Ugrunaaluk was a close relative of *Edmontosaurus*, found further south, and indeed was thought to be a variant of that species until work by Druckenmiller, Erickson and Hirotsugu Mori showed there were small but distinct differences in the bones of the skull. They argued that it was a unique animal, worthy of its own name, and published their findings describing *Ugrunaaluk* in 2015.[5] 'Because many of the bones from our Alaskan species were from younger individuals, a challenge of this study was figuring out if the differences with other hadrosaurs was just because they were young, or if they were really a different species', Druckenmiller told reporters. 'Fortunately, we also had bones from older animals that helped us realize *Ugrunaaluk* was a totally new animal.'[6] The team worked with Ronald Brower, an indigenous Iñupiaq speaker at the UAF's Alaska Native Language Center, to come up with a name that honoured the region's first people.

The remains of *Ugrunaaluk* are the most common fossils in the Liscomb Bone Bed, but other beds and nearby sites preserve additional material too: *Pachyrhinosaurus*, numerous teeth shed by tyrannosaurs and troodontids that were feeding on the herbivores, a dome-headed pachycephalosaur named *Alaskacephale* in 2006, mammal teeth and fish remains. Determining what is what, however, and deciding whether the remains represent new species is a significant challenge. 'What we're finding up here looks like someone took a zoo and put it through a [food processor]', Erickson says. 'Everything is broken up, so it's really difficult to say what we have. We just have to keep digging until we find diagnostic bones that tell us if it's a new species or not.'

One thing that is becoming clear is that many of the dinosaurs in this area are slightly different from their relatives down south. Erickson and Druckenmiller argue that they represent a unique northern fauna. In the Late Cretaceous, a vast inland sea ran the length of North America from the Arctic Ocean down to the Gulf of Mexico, creating two separate landmasses: Laramidia to the west and Appalachia to the east, both home to their own diverse faunas. The dinosaurs in Alaska lived at the northern extremity of Laramidia.

The types of dinosaurs found all the way down Laramidia are similar, and for a long time it was thought they represented a number of single species that were very widespread – dinosaurs such as *Edmontosaurus*, which have been found all the way from Alaska down through Alberta and Montana to Mexico. But in recent years, palaeontologists have been debating whether the dinosaurs found in each region are in fact endemic species that are similar to, but isolated from, the species in neighbouring regions.

According to Druckenmiller, looking at minute details of the bones from the far north, they see that many of them have distinctive features. 'When we looked at the fine details of the bones of *Ugrunaaluk* we realised that though it does resemble *Edmontosaurus*, it's not the same thing. So we felt justified in giving it a new name. And when we start looking at the same level of detail in the dinosaurs from Alaska, we realise that none of them are the same species as Alberta or Montana ... It's a lost world of dinosaurs in the Arctic that no one really knew about.'

The pair are currently working on material that includes new species to be described and reveals whole groups of dinosaurs not previously found in Alaska. One new species is a lambeosaur, a kind of crested duck-billed dinosaur similar to Russia's *Orolotitan* (see the previous chapter).

The Liscomb and other bone beds of the Prince Creek Formation have been excavated for many years, and people thought there wasn't much left to discover there, but these scientists have developed new methods to probe it for fossils and a new methodology for studying existing bones. 'We realised you have to learn to look in new ways, and since we figured that out we've been making a lot of really cool new discoveries', Druckenmiller says. 'We're describing new types of dinosaur, from bones and teeth, that we had no idea even lived in Alaska or the Arctic. As we get more material from these sites ... it's going to open up the story of what Arctic dinosaurs were really like.'

Extreme palaeontology

Fossil hunting in Alaska is expensive, difficult and dangerous. On the North Slope there's just a brief, three-week window to work each year. The river is clear of ice by May, but the scientists are not allowed to dig the bone beds on the Colville River until August, so as not to disturb peregrine falcons nesting on the cliffs. This explains why it can take a long time for new Alaskan discoveries to be made – not only are the fossils fragmentary and difficult to work with, but there is only so much material that can be collected in the brief field season and then carried out by small aircraft.

Erickson describes it as 'some of the most arduous dinosaur palaeontology that can be done'. The cold, rain, fog and snow are just the start of it. The researchers also have to contend with ubiquitous biting insects, and run the risk of encountering wolves and bears. Tents have been trampled by grizzlies, and with climate change there's even a chance of running into a polar bear. Simply getting out to field sites can be challenging. They fly in on small turboprop planes that land on gravel bars, but if the conditions aren't right they can't get in. The same is true when they have to leave, and on occasion they've been stuck for days waiting for the right weather window for a plane to get in.

The scariest thing is when the cliffs around the Colville River collapse without warning. These 20-metre-high bluffs are a mixture of unconsolidated sand and mud held together by ice. On warm days the sun melts some of the ice and 'without warning 100 tonnes of rock can come down on your head and there's nowhere to go', says Druckenmiller. 'You would get swept straight from the outcrop right down into the river. You could go from having a great day to being dead underwater and never discovered again in a matter of seconds.' There are only so many precautions the palaeontologists can take, and Druckenmiller says his biggest fear is that somebody will get hurt by a cliff coming down. 'When you have a warm few days ... all you hear throughout the night, maybe every 20–30 minutes, is the sound of a cliff collapsing somewhere within a mile or two of the camp.'

Despite the hardships, the participants look forward to the expeditions. Erickson says he enjoys speaking with the local Iñupiaq people, with whom he and Druckenmiller have developed a good rapport, and the pair visit them at local villages and schools to talk about their discoveries. 'It's really fun ... I admire these people. What a tough place to live ... We meet with their elders

Myth buster. *Above* Known only from a pair of terror-inducing, 2.4-metre-long fossil arms for nearly 50 years, 6-tonne *Deinocheirus* was finally revealed in 2014 to be a weird member of the omnivorous ornithomimid group. It lived in Mongolia 70 million years ago. SOURCE Andrey Atuchin

Bite force. *Below* Author John Pickrell discovered the partial lower jaw (with teeth attached) of a tyrannosaur relative called *Alectrosaurus* in Mongolia's Gobi Desert in 2015. Pickrell was hosting a fossil dig that was a collaboration of *Australian Geographic* and the Mongolian Academy of Sciences. SOURCE John Pickrell

Lightning Claw. *Above* Yet to be officially described as a species, this slender, 7-metre-long megaraptorid prowled along the swamps and floodplains of Lightning Ridge, New South Wales, 110 million years ago. Today part of outback Australia, this region would have been near the Antarctic Circle during the Cretaceous. SOURCE Julius Csotonyi

Outback gems. *Below* A selection of remarkable opalised fossils from Lightning Ridge, Australia. Top: A 'yabby button' from the head of a freshwater crayfish; beautiful purple pine cone; crocodile tooth in valuable black opal with red colour; theropod tooth, possibly a megaraptorid. Bottom: Pine cone; snail shell; turtle vertebra, sauropod tooth, likely a titanosaur. SOURCE Robert A. Smith/Australian Opal Centre

Denizens of Dinosaur Cove. *Above* Three-metre-long tyrannosaur *Timimus hermani* and large-eyed ornithopod *Leaeallynasaura amicagraphica*, found in Victoria in south-eastern Australia, were animals that lived within the Antarctic Circle and had adaptations to cope with the cold and dark winter. SOURCE Lida Xing/*Australian Geographic*

Flash mob. *Below* Ceratopsians, such as *Regaliceratops*, likely used their horns and frills to attract mates and display to members of their own species. In life these may have been brightly coloured and distinctively patterned. SOURCE Royal Tyrrell Museum of Paleontology/Julius Csotonyi

Centrosaurinae
1. Xenoceratops foremostensis
2. Sinoceratops zhuchengensis
3. Pachyrhinosaurus lakustai
4. Pachyrhinosaurus perotorum
5. Pachyrhinosaurus canadensis
6. Achelousaurus horneri
7. Einiosaurus procurvicornis
8. Monoclonius lowei
9. Coronosaurus brinkmani
10. Nasutoceratops titusi
11. Styracosaurus albertensis
12. Rubeosaurus ovatus
13. Centrosaurus apertus
14. Spinops sternbergorum
15. Avaceratops lammersi
16. Diabloceratops eatoni
17. Albertaceratops nesmoi
Ceratopsoidea
18. Turanoceratops tardabilis
19. Zuniceratops christopheri
Protoceratopsidae
20. Protoceratops andrewsi
21. Protoceratops hellenikorhinus

Chasmosaurinae
22. Anchiceratops ornatus
23. Coahilaceratops magnacuerna
24. Kosmoceratops richardsoni
25. Agujaceratops mariscalensis
26. Vagaceratops irvinensis
27. Mojoceratops perifania
28. Medusaceratops lokii
29. Arrhinoceratops brachyops
30. Chasmosaurus belli
31. Chasmosaurus russelli
32. Pentaceratops sternbergi
33. Utahceratops gettyi
34. Eotriceratops xerinsularis
35. Triceratops horridus
36. "Yoshi's Trike"
37. Judiceratops tigris
38. Titanoceratops ouranos
39. Triceratops prorsus
40. Torosaurus latus
41. Bravoceratops polyphemus

Weird and wonderful. *Above top and right* An incredible number of new ceratopsian horned dinosaurs has been found in the last 15 years, the majority in western North America. These have revealed a remarkable array of styles of horns and frills, as illustrated here by Canadian palaeoillustrator Julius Csotonyi. SOURCE Julius Csotonyi

Crowning glory. *Below right* Impressively ornamented *Regaliceratops* was for a long time, known simply as 'Hellboy' a reference to both the nightmare of extracting this fossil skull from the rock and its unusual headgear. *Regaliceratops peterhewsi* means 'royal horned face, named in honour of Peter Hews'. SOURCE Royal Tyrrell Museum of Paleontology

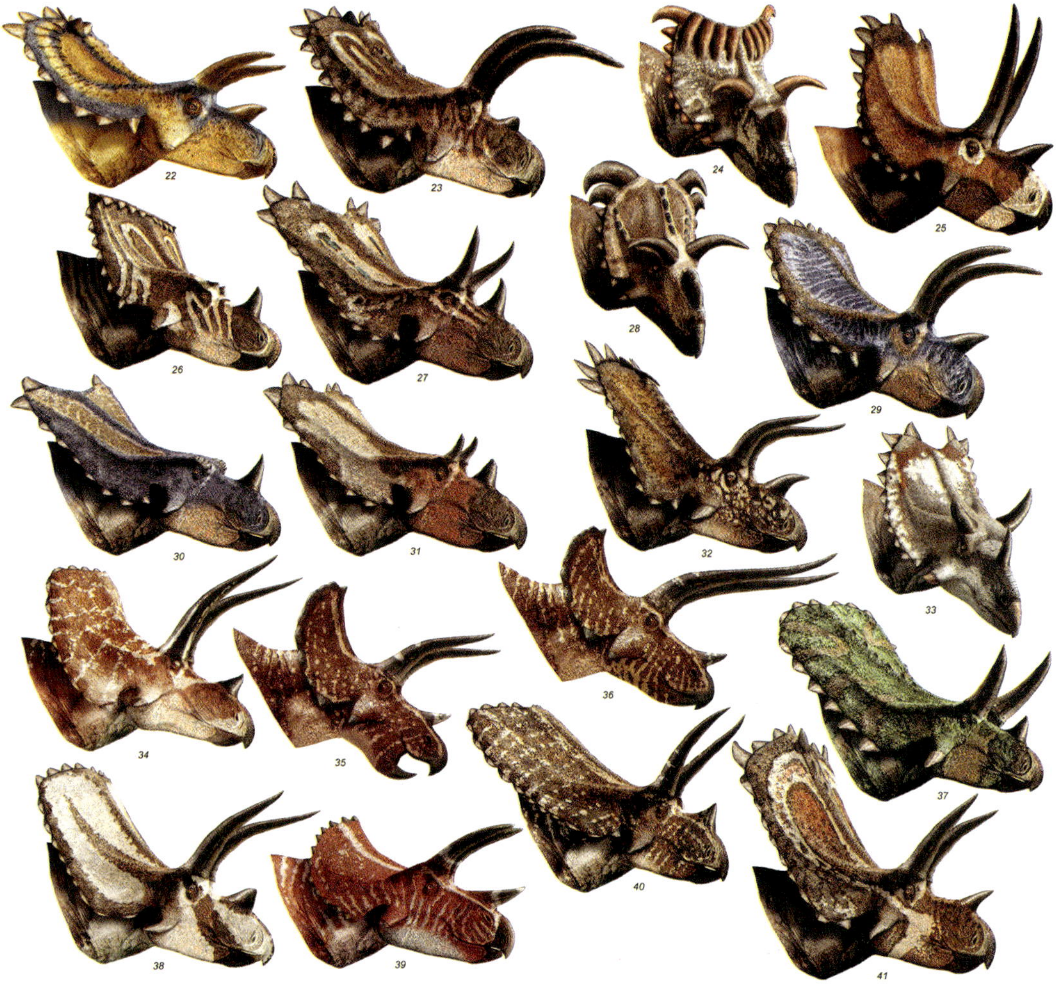
22
23
24
25
26
27
28
29
30
31
32
33
34
35
36
37
38
39
40
41

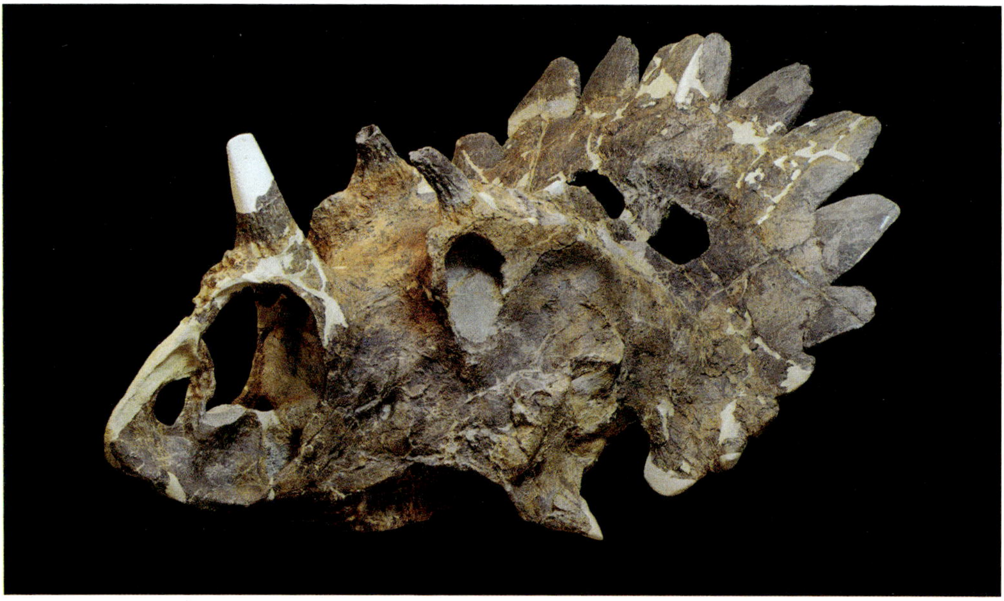

Plumage power. *Above* Fossils of herbivorous ornithischian *Kulindadromeus*, from the Jurassic of Siberia, provided evidence in 2014 that feathers were not restricted to the carnivorous theropods but may in fact have been a trait shared by all groups of dinosaur. SOURCE Andrey Atuchin

Northern grazer. *Top right Ugrunaaluk* (oo-GREW-nah-luk) *kuukpikensis* – found in great beds of jumbled fossil bones at Alaska's North Slope – was originally thought to be *Edmontosaurus*. But increasing evidence is showing that this hadrosaur and many other animals here were part of a unique Arctic province of Cretaceous dinosaurs. SOURCE James Havens

Cool customers. *Below right* Ceratopsian *Pachyrhinosaurus perotorum* and Cretaceous bird *Gruipeda vegrandiunis* are two of the animals now known to have lived in the great polar forests of Alaska 70 million years ago. Tracks of these animals have been found in Denali National Park in the south of the state. SOURCE James Havens/Wikimedia

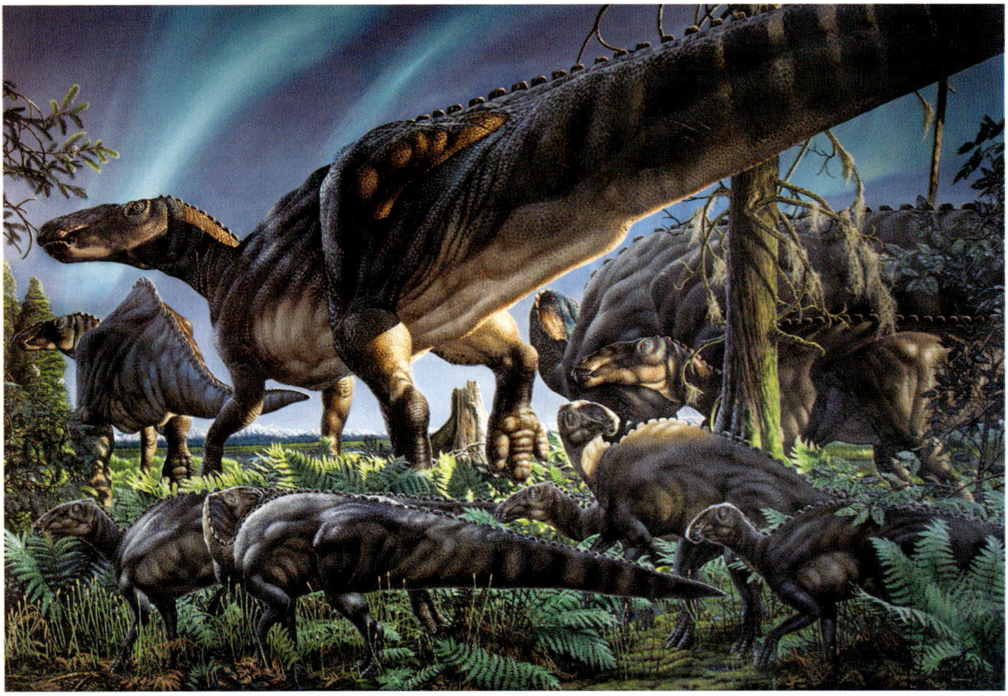

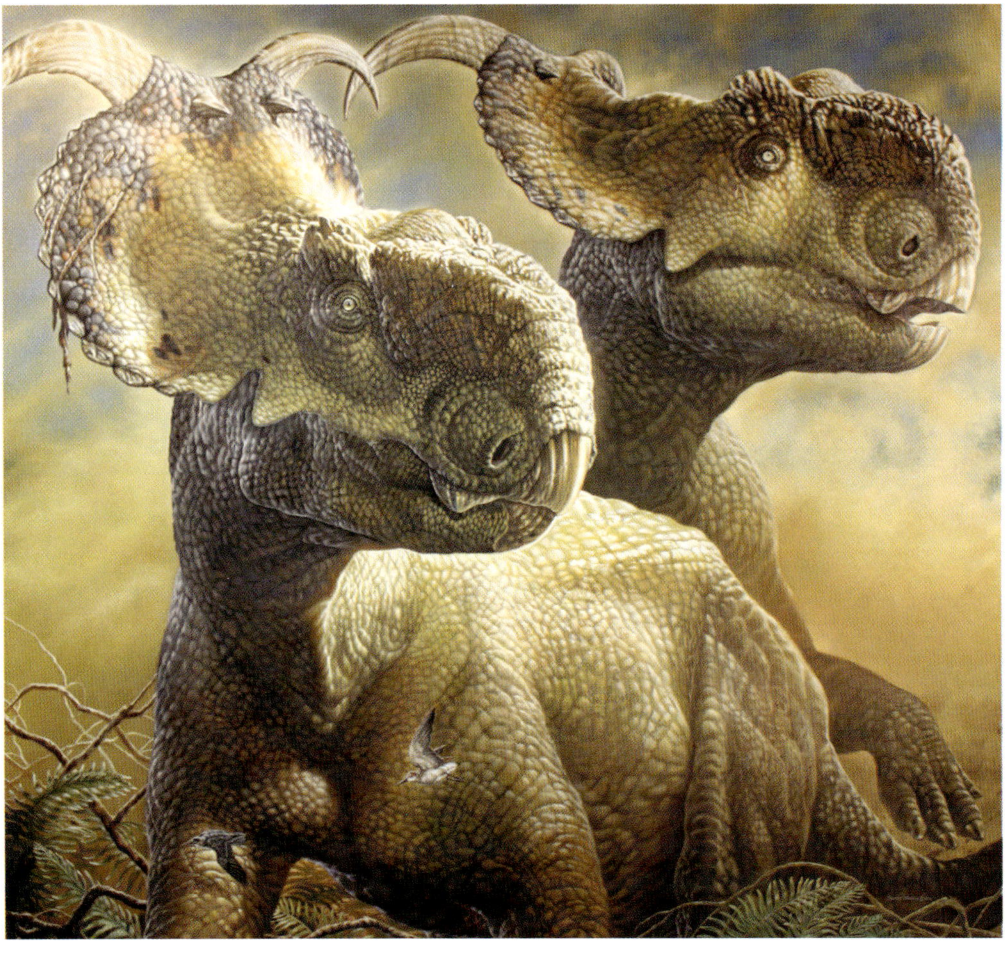

Dinosaur Park Formation. A beautiful scene of Late Cretaceous dinosaur species found at Dinosaur Provincial Park in Alberta, Canada, illustrated by Julius Csotonyi. From left: *Chasmosaurus, Lambeosaurus, Styracosaurus, Scolosaurus, Prosaurolophus, Panoplosaurus* (and a herd of *Styracosaurus* in the background). SOURCE Julius Csotonyi / JC Mallon et al.[1]

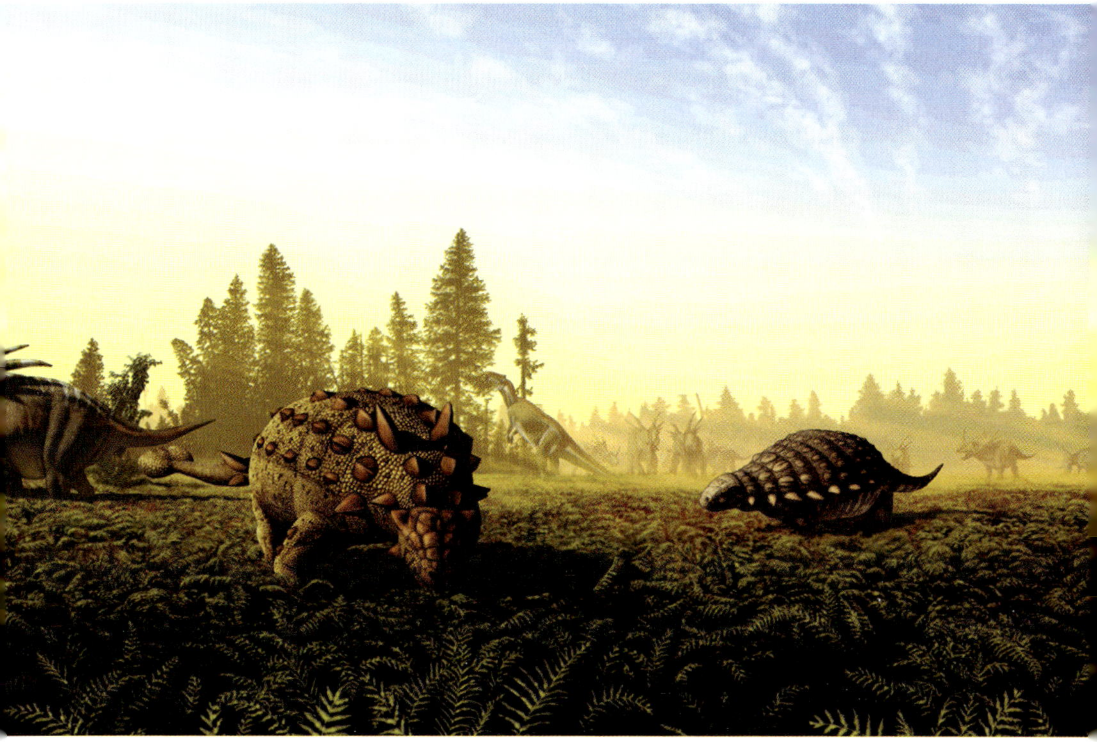

Prehistoric songstress. Palaeoillustrator Emily Willoughby takes her cues from modern birds when illustrating theropod dinosaurs such as *Balaur bondoc*, here pictured in the Late Cretaceous environment of Haţeg Island (today the Romanian region of Transylvania). SOURCE Emily Willoughby

Hunger games. *Above* Enormous azhdarchid pterosaurs, such as *Hatzegopteryx* and *Quetzalcoatlus*, may have stood as tall as a giraffe and snacked upon young sauropods that were yet to reach their titanic potential. SOURCE Mark Witton and Darren Naish (2008)[1]

Raptor reality. *Below* Feathered reconstructions of dinosaurs are proliferating in museums across the world, such as this model of a Velociraptor, which was part of the Tyrannosaurs: Meet the Family exhibit shown at the Australian Museum, Sydney. SOURCE Australian Museum/James Morgan

River giant. *Above* Prowling the waterways of North Africa 100 million years ago, giant semi-aquatic predator *Spinosaurus* was even larger than *T. rex*; 15 metres or more from head to tail, it had a 2-metre-tall sail on its back. Though it may sometimes have hunted dinosaurs, the huge fish with which it shared its environment are thought to have been its key prey. SOURCE Davide Bonadonna

Frozen crested lizard. *Below* One of the best known early Jurassic carnivores, *Cryolophosaurus ellioti*, was named in honour of David Elliot, the geologist who found it 640 kilometres from the South Pole and 4000 kilometres up Mt Kirkpatrick in the Transantarctic Mountains. This reconstruction is at Royal Ontario Museum in Toronto. SOURCE D. Gordon E. Robertson/Wikimedia

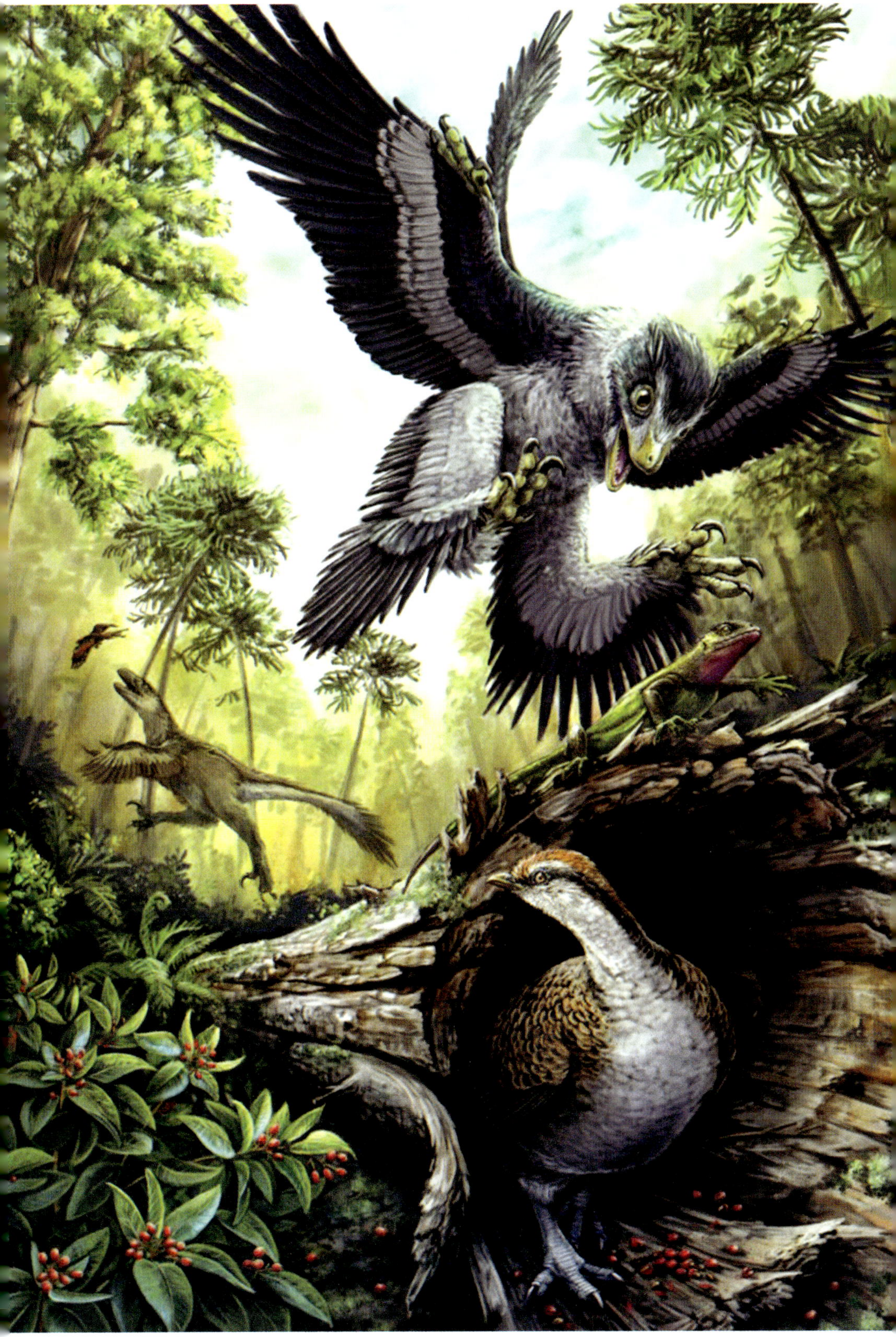

Lucky break. *Previous page* This scene depicts the life of Canada's Hell Creek Formation 66 million years ago. Research from scientists David Evans, Caleb Brown and Derek Larson suggests that birds may have made it through the mass extinction that killed other dinosaurs because their beaks uniquely equipped them to eat seeds – one food source that remained in the wake of the cataclysm. Small feathered theropods with teeth, including this four-winged flying dromaeosaur (front), became extinct at this time. SOURCE Royal Ontario Museum/ Danielle Dufault

Winged wonder. *Above* Zhenyuanlong was a 2-metre-long Chinese relative of *Velociraptor*. Its fossil, revealed to the world in 2015, showed the clear impressions of small wings with large feathers, adding to the evidence that most dromaeosaurs, *Velociraptor* included, had small wings. SOURCE Emily Willoughby

Chicken from hell. *Top right* Reconstruction of the skeleton of the oviraptorid *Anzu wyliei* at the Rocky Mountain Dinosaur Resource Center in Woodland Park, Colorado. Described in 2014, this incredibly bird-like, beaked species was named for a feathered demon in Mesopotamian mythology. SOURCE MCDinosaurhunter/ Wikimedia

Gondwanan cannibal. *Below right* *Majungasaurus crenatissimus* was the first dinosaur to be found on Madagascar, by a French soldier in 1896, but a more complete specimen would not be found for another century. This 7-metre-long abelisaur had a skull with a bumpy, sculpted texture; a small, domed horn; and absurdly small forelimbs, with reduced digits. SOURCE Deviant Paleoart/Wikimedia

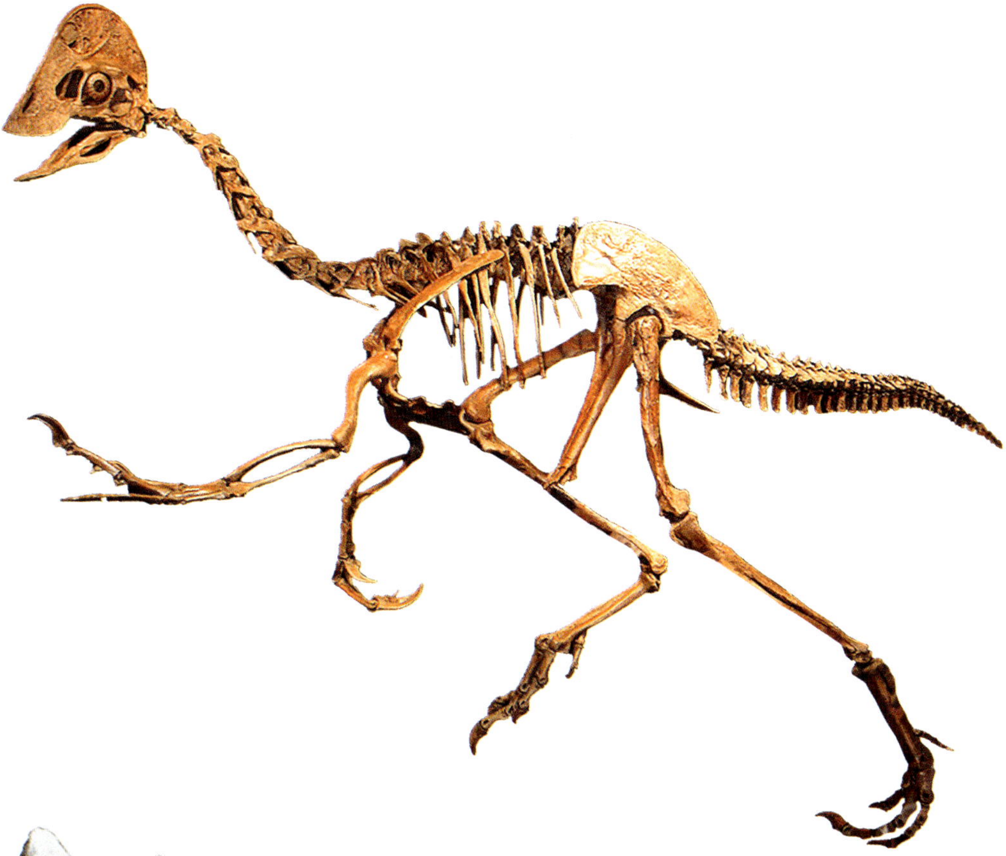

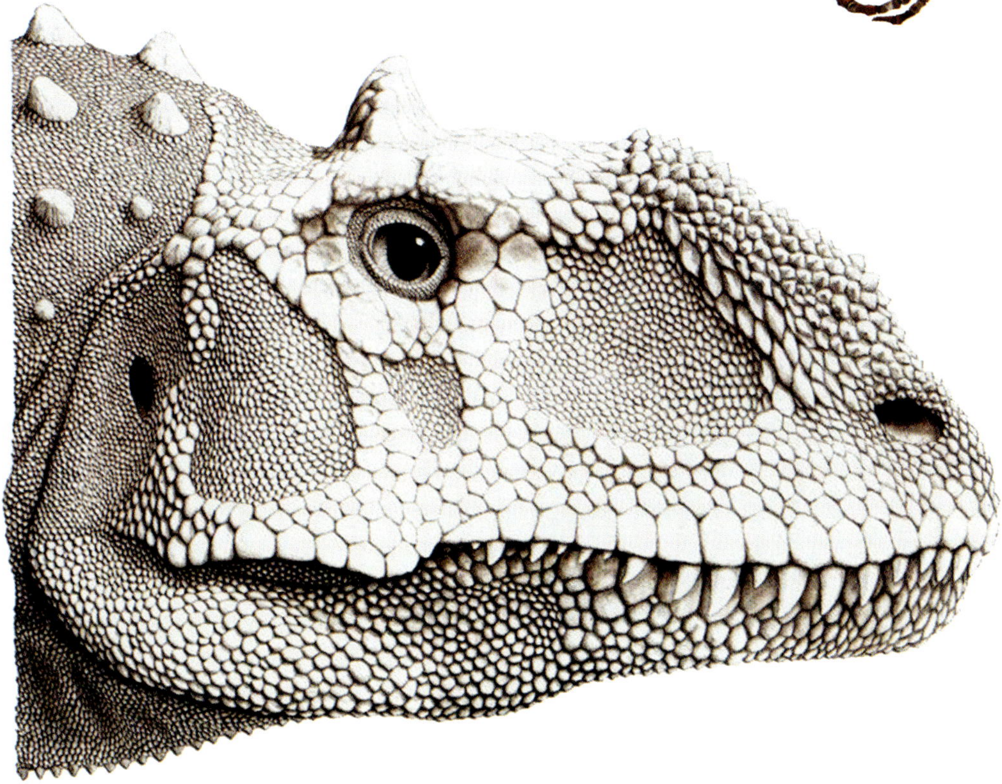

Giant mystery. Sauropods – such as (from left) *Camarasaurus*, *Brachiosaurus*, *Giraffatitan* and *Euhelopus* – included the largest land animals that ever lived, some of which may have reached lengths of 38 metres or more and weighed 70 tonnes. Why they grew so large is still to be resolved. SOURCE dmitrchel@mail.ru/Wikimedia

Caped crusader. This astonishing, 60-centimetre-long dinosaur left experts stunned when it was revealed in 2015. From Hebei in China, it had membranous wings akin to those of a bat, as well as a body covered in downy fluff, and ribbon-like feathers on its tail. *Yi qi* means 'strange wing'. SOURCE Emily Willoughby

1   JC Mallon & JS Anderson, 'Skull ecomorphology of megaherbivorous dinosaurs from the Dinosaur Park Formation (Upper Campanian) of Alberta, Canada', PLOS ONE, vol. 8, no. 7, article no. e67182.
2   M Witton & D Naish, 'A reappraisal of azhdarchid pterosaur functional morphology and paleoecology', PLOS ONE, vol. 3, no. 5, article no. e2271.

and try to honour what they might call these animals, and they appreciate that we're doing that.'

The pygmy polar *T. rex*

Fiorillo manages Alaska's other long-running dinosaur research program and, along with his colleagues at the Perot Museum in Dallas, has been coming here nearly every year since 1998. One of his most exciting discoveries, a small tyrannosaur called *Nanuqsaurus*, was made almost by accident.

Four out of five bone beds around the Colville River are hadrosaur, but in 2006 Fiorillo was working on something else. At a 4-metre-square site called the Kikak-Tegoseak Quarry, he was excavating the skull of a *Pachyrhinosaurus*, the only other dinosaur that appears often in these deposits. Slowly, as they excavated the site, they realised the quarry contained the skulls of not just one, but 10 *Pachyrhinosaurus* – the remains of animals that had likely been killed in a flash flood. 'We knew that this quarry had a number of horned dinosaurs, and that was unusual given that most of the bone beds here had hadrosaur bones', Fiorillo says. 'This was something different and that's what drew my attention.'

Over two years, his team excavated 6 tonnes of fossil-bearing rocks from this quarry. These were lifted out by helicopter, flown out from Fairbanks on a fixed-wing aircraft and then trucked to Dallas. At around this time, the Perot Museum was in the throes of putting together exhibitions for its new building. 'I recognised that we had in the excavation at least one skull of a *Pachyrhinosaurus*, and it was potentially a beautiful skull and display-worthy, so really our focus was on getting that skull ready for display in our new dinosaur hall.'

When Fiorillo talks about one skull, what he means is that there were large numbers of fragments of one skull that had to be pieced back together. The ice, snow and permafrost, and cycles of freezing and thawing, tend to break the fossils into pieces. 'It actually took about four years of preparation and reconstruction to put it together so it would be stable enough to go on display', he says. Another few years' work after that revealed this was a new species, different from the two other known *Pachyrhinosaurus* species found further south.

Until the publication of that finding in 2012 and then a juvenile specimen in 2013,[7] all their attention had been focused on the horned dinosaur fossils, and they didn't even think about what else might be jumbled in with the remains. Once they'd passed that hurdle, however, Fiorillo and Perot Museum fossil preparator Ronald Tykoski decided to look at some of the other, non-ceratopsian bones they'd noticed sticking out of the blocks of rock brought from the quarry. A bit of probing and they found a jaw with sharp, pointy teeth and then other parts of a skull. They decided this was part of a tyrannosaur, but the pieces were very puzzling. The teeth were tiny, suggesting it was a juvenile, but other features were only usually found in adult dinosaurs. Eventually they realised they had a very small adult tyrannosaur.

T. rex, which was a close relative, was 12 metres long and weighed about 7 tonnes, but the species Fiorillo described in 2014 as *Nanuqsaurus* was just 7 metres long and weighed as little as 1 tonne.[8] 'During the course of our excavation we were also extracting bones with teeth marks in them and scratching our heads wondering who was eating our *Pachyrhinosaurus*', Fiorillo says. 'Well, once we had *Nanuqsaurus*, we were able to answer that question.'

The size of the carnivorous dinosaurs from Alaska is puzzling. While *Nanuqsaurus* is smaller than its relatives, another theropod found there, *Troodon*, appears to be larger than its southern relatives. There is a loose biological trend called Bergmann's Rule whereby many animals are larger in populations closer to the poles. But not everything follows this trend, Fiorillo says, pointing to grizzly bears in Alaska, which tend to vary in size based more on where they can exploit salmon as a food resource.

The reduced size of *Nanuqsaurus* may have more to do with

insular dwarfism – the phenomenon that led to the pygmy dinosaurs found by Baron Nopcsa in Transylvania. Of course Alaska wasn't adrift in the ocean in the Late Cretaceous, but it was relatively isolated from more southerly Laramidia by the Brooks Range of mountains, which Fiorillo argues effectively turned it into an ecological island. A restriction on the food resources available here, exacerbated by the 3–4 months of winter darkness, may have meant smaller size was an adaptive advantage for an apex predator.

Troodon, a wolf-like carnivorous dinosaur akin to *Velociraptor*, had a large sickle claw on each foot and was found throughout much of Laramidia. It's fairly rare in Montana, Alberta and Texas, but in Alaska the presence of its distinctive teeth suggests it was the most common predatory dinosaur by a long shot. Not enough skeletal material has yet been found to confirm whether this is a new species or simply a larger version of the same one found elsewhere. 'The teeth are significantly larger than the *Troodon* found further south ... about 50 per cent larger', says Fiorillo, who says the Alaskan animals would have been 3.5 metres long. 'One of the things that makes *Troodon* unique is that it has the largest eye-orbit-to-body-size ratio of any dinosaur ... We argue that *Troodon* was pre-adapted for the light conditions of the Arctic, which allowed it to be so abundant.'

The large size of *Troodon*'s eyes may have meant that in the Arctic it was able to outcompete other kinds of predator that were more common elsewhere. Fiorillo points to modern ecosystems where wolves and coyotes once co-existed but the wolf has been removed. 'The coyote's response is that they grow bigger. The top predator is removed, so they can expand into their niche space. There's a modern basis for saying that's what we are also seeing in *Troodon*.'

Forensic zoology and the Bering Land Bridge

Gregory Erickson is a palaeobiologist or, as he says, 'a zoologist doing forensic science' – he tries to answer questions about dinosaur biology using mere crumbs of information. Erickson is interested in how dinosaurs lived, how they grew and developed, and what their ecosystems were like.

Though he's now based in Florida, Alaska is the state of his birth, and he had followed the dinosaur discoveries up there with interest. When Druckenmiller moved to the University of Alaska Museum of the North in Fairbanks and began to realise there were unfamiliar dinosaur bones in the collection, he decided to build an interdisciplinary team to study them, find new fossils and answer novel questions. Erickson, an expert on dinosaur growth rates, was an obvious choice to join him in a collaboration they dubbed the Paleo-Arctic Research Consortium (PARC).

'I got interested in using the animals up there as a natural experiment. We started to look at palaeobiological questions about physiology and migration', says Erickson who began working with Druckenmiller in about 2008. The fact that no modern cold-blooded (ectothermic) species, such as frogs, lizards or crocodiles, are found as fossils at the North Slope suggests that dinosaurs had a fundamentally different metabolism from these creatures.

In the 1980s, many palaeontologists still thought of dinosaurs as typical reptiles, but a lot has changed since then. For a start, we know that avian dinosaurs – birds – are warm-blooded, which has led many to surmise that all theropods shared this endothermic physiology too. Erickson believes that all dinosaurs, including the herbivores, had some kind of endothermic metabolism, which was at the low end of the spectrum for growth rates, perhaps similar to slow-growing mammals today. 'I've done a lot of research on dinosaur growth rates, which suggests they grew at the low end of the endotherm [warm-blooded] range', he says. 'But homing in on that is hard when you only have bones to work with. It's difficult to put numbers on.' Finding more varieties of dinosaur in Alaska and examining bone slices for growth rings may help to answer these questions.

'I suspect they were very good at controlling their metabolic rates, because it doesn't seem that they were migratory', Fiorillo agrees. 'If they're not migrating then they are going through long periods of light and dark. So, much like animals today, which change their diet and metabolic needs over the course of a season, I suspect you would see that reflected somehow within these fossils.' In an attempt to test this idea he looked at the wear patterns of hadrosaur teeth from Alaska down to Mexico, in the hope of seeing differences that might indicate the animals were foraging on different kinds of plants. The results were inconclusive, so the next step, as yet not done, is to analyse the tooth enamel chemically to test for different foodstuffs.

If the Alaskan theropods were warm-blooded, it seems likely some were feathery animals with fuzzy coats of plumage for insulation. While the bone beds of the North Slope are the wrong place to find fossilised feathers, Fiorillo says the fine-grained lake deposits of some of the southern footprint sites may one day offer up a feathered dinosaur. 'Given the discoveries in China, it would be my expectation that some day we will find feathered dinosaur remains in Alaska.'

One other major puzzle to solve is how these dinosaurs ended up in Alaska and whether it acted as a stepping stone for dinosaurs occasionally passing between Asia and North America. 'The therizinosaur tracks at Denali show us something about the Asian ecosystem encroaching onto North America', Fiorillo says. 'So even though we don't have exactly the same dinosaurs, there's something about the ecosystem structure that we see in Asia that's made it at least as far as Alaska in North America … Alaska represents that gateway for communication between Asia and North America. So if we can study the timings of these things in Alaska, we'll know about the dinosaurs in Alberta and the Gobi too.'

There has long been an idea of faunal exchange between Asia and North America during the Cretaceous, via the Bering Land Bridge, and many of the types of species – hadrosaurs, tyrannosaurs, ankylosaurs – are similar on both continents. Furthermore, the ancestral forms of the dinosaurs found in America all appear to be Asian animals, creatures such as the early tyrannosaurs *Dilong* and *Guanlong*, and early ceratopsians *Protoceratops* and *Yinlong*.

Another question that has long been burning in many palaeontologists' minds is why large ceratopsian horned dinosaurs are almost absent in Asia when they were diverse and abundant in North America. For a long time they were thought to be completely absent, until a team led by Xu Xing discovered the first, a 6-metre-long horned species called *Sinoceratops* in 2010.[9] The best explanation so far is that there was some kind of ecological filter or barrier preventing large horned dinosaur species from making the journey into Asia. Fiorillo tends to believe that it wasn't a water barrier, but also that just because there was land doesn't mean these dinosaurs could cross it – they have to have food and shelter along the way. 'Was there a food source that climate didn't allow to be in a place that the ceratopsians needed it to be to continue the migration back into Asia?' he asks. 'I don't know the answer ... That ecological barrier is something more subtle that still needs to be teased out of the ecological record.'

The final frontier

Alaska is gigantic and still largely unexplored by fossil hunters. Many regions have rocks of the right age and should contain fossils but are yet to be prospected. The PARC team has begun to make guesses about where these might be, which is what led them to the Yukon River in 2013. 'Ten years ago there were really just two sites, but now we're realising there's a lot of potential and a lot of new sites', Druckenmiller says. To access these you need to know how to work in really remote landscapes.

The team has a number of discoveries to announce, including varieties of dinosaur never found in Alaska before. 'Given that we are still at the stage of recognising whole new groups tells us that we are missing some pretty big gaps', he says. 'We need to get out in the field more and do as much collecting as possible. It would be really nice to find some sauropod bones as well as more types of tracks.'

For the moment, Fiorillo is focusing on preparing additional fossils of *Pachyrhinosaurus* and hadrosaurs, which will increase sample sizes, allowing his Perot Museum team to further understand the physiology and morphology of these animals. He is also continuing to prepare the 6 tonnes of rock brought back from the Kikak-Tegoseak Quarry, which yielded the two new species of dinosaur. He says that while new species will be found, there might eventually be a limit, because the biodiversity of polar regions was likely to have been low in the Cretaceous, just as it is today.

The Alaska Peninsula, which juts down towards the Aleutian Islands and Russia's Kamchatka Peninsula beyond, is another remote region where dinosaur footprints have recently been found. The earliest recorded evidence of dinosaurs in Alaska, as far as Fiorillo can tell, is a photograph of a footprint from a 1930s French encyclopaedia, which was probably taken on the peninsula.

Once they have a better handle on the species that are in Alaska and the associated islands, the scientists can one day start asking questions about how similar these species are to polar dinosaurs found in Australia or Antarctica, and whether they had the same kinds of growth rates and adaptations. The sites in Australia are pretty tough going, says Erickson. 'The dinosaur-bearing rocks they are dealing with are few and far between, difficult to access and much harder; they chip and chip and chip and they are lucky to get something each summer. Whereas, literally, we could dig up a thousand bones in a summer if we wanted to. Alaska is basically the last frontier', he adds. 'It's virtually unexplored in terms of vertebrate palaeontology, so we think we're going to find a lot of new species.'

Despite the difficulties facing dinosaur hunters in Australia, many important discoveries have been made, at a number of sites. Some of the most exciting new dinosaur finds are the opalised fossils coming from Lightning Ridge.

8

THE HIDDEN TREASURES DOWN UNDER

LIGHTNING RIDGE, AUSTRALIA

Buried in the outback are the fossils of prehistoric creatures cast in precious stones. It seems almost beyond belief, but the dinosaurs of Lightning Ridge are sometimes preserved as precious opal, a rare gem that diffracts and refracts light into beautiful shimmering colours.

A thunderstorm crackles on the horizon in forests not far from the edge of the Eromanga Sea. It is 110 million years agao, during the Cretaceous Period, and this vast body of water divides Australia, which is much further south than it is today; still part of the supercontinent Gondwana, it straddles the Antarctic Circle. Despite lengthy periods of winter darkness at this high latitude, large evergreen conifers including hoop and kauri pines dominate the dense polar forest, alongside seed ferns, horsetail ferns and cycads.

The foliage rustles and branches snap as something moves through the shadows beneath a blanket of dark, forbidding cloud. As the wind picks up and heavy rain begins to fall, there is a deafening clap of thunder, and a flash of light briefly illuminates a medium-sized, slender carnivore prowling these seasonal coastal forests and waterways. This 7-metre-long dinosaur, less bulky than a *T. rex* or an *Allosaurus*, has relatively long arms with huge claws. It has a short downy covering of feathers that glistens with moisture. Emitting a low rumbling call, it shakes itself, fluffs up its plumage and nestles into a hollow beneath a conifer, where it tucks its head under its arm in the same roosting position as a modern bird, and waits for the conditions to improve.

This dinosaur, dubbed Lightning Claw, is known from some very unusual fossils unearthed by miners in an outback town in northern New South Wales. A dusty settlement with a frontier spirit, Lightning Ridge is an eight-hour drive from Sydney. It's the only significant dinosaur site in the state and one of the richest in Australia. Revealed in 2015 but not yet formally named,[1] Lightning Claw is one of a growing number of megaraptorid carnivores that diversified and spread across Gondwana.

These include *Megaraptor*, discovered in Patagonia in 1998, and further Argentinean species *Orkoraptor* and *Aerosteon*, described in 2008 and 2009. Australian occurrences in addition to Lightning Claw include *Rapator*, described in 1932 from a single bone from

Lightning Ridge; *Australovenator*, a much more complete, 5-metre-long dinosaur from Winton, Queensland, described in 2009; and an arm bone found at Dinosaur Cove in Victoria. Questions remain about possible Northern Hemisphere megaraptorids (*Fukuiraptor* from Japan, *Eotyrannus* from the United Kingdom and *Siats* from the United States) and whether this group of carnivores is most closely related to tyrannosaurs or allosaurs.

What we do know is that the dinosaurs that dominated the Northern Hemisphere are often quite different from those found in South America, Australia, Antarctica, Africa, the Arabian Peninsula, India and Madagascar – the landmasses that once formed Gondwana. In Australia, the fossil record is scant – more species of dinosaur have been found in a single quarry in China, than the 20 or so species that we know from Down Under after a century of searching – but the species turning up here with increasing frequency are relatives of those discovered in Patagonia over recent decades.

'It's been really interesting to discover an entire new family of carnivores', says Diego Pol of the Museo Paleontológico Egidio Feruglio (MEF) in Trelew, Argentina, who was involved in the Patagonian megaraptorid finds. 'My guess is that we will find a lot more lineages in Australia and Antarctica that are present also in South America ... and we may have some surprises in the future too.'

Hard-won outback fossils

Australian dinosaurs have often been thought of as primitive, relict species that diversified in other parts of the world and ended up, as if stranded, here on the periphery of Gondwana, but the discovery of Lightning Claw and a number of other new dinosaurs suggests that's a fallacy. At 110 million years old, Lightning Claw is the oldest known megaraptorid, and the isolated arm bone from Victoria also dates to the Early Cretaceous. Phil Bell, at the University of New England (UNE) in Armidale, New South Wales, who is studying the dinosaurs of Lightning Ridge, says the evidence may point to the group originating in Australia, then fanning out to colonise other parts of the supercontinent. Similarly, a primitive Australian ankylosaur called *Kunbarrasaurus*, described in 2015,[2] and a related species *Minmi*, discovered in 1964, also hint that these armoured dinosaurs may have evolved here before spreading out across the world. Could Australia therefore have been the centre of a series of radiations of Gondwanan dinosaurs?

The problem with answering this question is that the Australian fossil record from the Mesozoic (the geological era encompassing the Triassic, Jurassic and Cretaceous periods) is very poor, and until recently this has left us hamstrung. Antarctica (see chapter 11) is the only continent with a poorer fossil record, and that's only because it's encased in ice. In other parts of the world, such as Canada and Mongolia, dinosaur hunters have the most luck searching in upland regions with rapid erosion and little plant cover, where fossils are exposed and easily spotted. But Australia has some of the oldest rocks on earth, and much of it is very ancient, flat and heavily weathered, with little recent geological activity and few outcrops of dinosaur-aged rocks.

A series of exciting fossil finds in four different areas of Australia – using four different methodologies to access them – has, however, resulted in much new information, and there are many more Australian specimens waiting to be described, which could double the number of known species from Down Under within a number of years.

In the opal fields of New South Wales, South Australia and Queensland, miners often turn up fossils of giant marine reptiles, dinosaurs, turtles, crocodiles and invertebrates. These typically small and isolated bones are preserved as opal – either dull common opal or 'potch', or precious opal, which exhibits a beautiful play of spectral colours and comes in a seemingly infinite variety of hues and patterns. Opal is one of the rarest precious gemstones, and although it's found in a number of countries, Australia has historically produced the bulk of gem-quality stones. The conditions that lead to opal formation are poorly understood, but in some locations they involve silica-rich groundwater in fine clays with high organic content, says Elizabeth Smith, a palaeontologist at the Australian Opal Centre (AOC) at Lightning Ridge and world expert on the region's fossils. Since the 1970s, she and her husband Bob have been sifting through mine tailings, liaising with opal miners and studying fossils to reveal much of what we know about these unusual remains.

It's thought that opal forms as a silica gel that concentrates in subterranean fissures and cavities, eventually hardening to form mostly common opal that lacks the sparkle of its rarer sibling. Occasionally, the gel gets trapped in a cavity created by plant or animal remains and – like jelly in a mould – forms an opalised fossil. Sometimes the silica solution also permeates the bones. Though the fossils are typically small and rarely of gem quality, the results can be breathtaking. Lightning Ridge has yielded a diverse and beautiful array of fossil remains, ranging from striking turquoise pine cones and snail shells, to sparkling red plesiosaur teeth and shimmering blue dinosaur vertebrae. Fish such as sharks, and turtles, crocodiles, pterosaurs, birds and monotreme mammals are found here too.

More than 1000 kilometres north-west of Lightning Ridge, in central Queensland, a series of large dinosaur discoveries has been made in the last 10–15 years by fossil hunters from the Australian Age of Dinosaurs Museum in Winton and the Queensland Museum in Brisbane. To find fossil beds they first search for fossil fragments on the surface of the dusty outback plains. These small pieces, which look like conventional rocks, are exhumed naturally by an unusual 'convection' effect created by the wetting and drying of the clay-rich soil. When they find a candidate site, the palaeontologists clear away the top metre or so using a bulldozer. This allows them to reach bigger fossils below. Dinosaurs discovered here include *Australovenator*, the megaraptorid mentioned earlier, and several giant sauropods, such as *Diamantinasaurus*, related to the Patagonian titanosaurs.

At the third fossil-bearing locality, Dinosaur Cove, on the coast of Victoria in south-eastern Australia, the method is different again. Here, erosion from the sea has exposed dinosaur fossils encased in a strip of hard rock that was once an ancient stream or river channel. The going here is tough, and the fossils – such as the large-eyed ornithopod *Leaellynasaura* and the small carnivore *Timimus* – sometimes have to be excavated with the help of miners, then chiselled out of solid rock by palaeontologists led by Tom Rich at Museum Victoria in Melbourne.

Four thousand kilometres away on the opposite corner of the continent is the Kimberley region of northern Western Australia, another dinosaur-bearing area of new interest. Here, Steve Salisbury and his University of Queensland team have used aerial photography from drones, photogrammetry and local Aboriginal knowledge to find and study hundreds of footprints and trackways along rock shelves in the intertidal zone. These ichnofossils (fossils bearing traces of life, such as burrows or footprints, rather than bones) reveal a rich fauna of more than 20 types of dinosaur from 130 million years ago during the Early Cretaceous – and there is hope that fossilised bones may one day be discovered here too.

The ghosts of Gondwana

Elizabeth Smith arrived in Lightning Ridge in a campervan with her husband, Bob, in 1973. They never expected to stay, but the atmosphere of the Ridge and the 'crazy quest for opal' of this frontier country got into their blood, as it has with so many others. Within a few years they'd taken up a mining claim at Lunatic Hill on the town's Three Mile opal field – one of the world's richest. Their simple stone shack sat amid open shafts and mullock heaps, rusted machinery and gnarled bimble box, wild orange and whitewood trees, everything powdered with white dust. As Bob dug through the white clay 20 metres below on the hunt for valuable 'nobbies' of black opal, Smith began to poke about in the mullock heaps of dirt at ground level, some of which had been brought to the surface by miners more than 50 years before.

She began to find small and sometimes beautiful scraps of opal sculpted into strange shapes. Having heard that fossils were occasionally found here, Smith, who had a love of natural history, began to question what the objects might be. 'On hands and knees, nose to the opal dirt, I was enchanted. There were shreds of forests and fragments of long-extinct animals', she writes in her book *Black Opal Fossils of Lightning Ridge*. 'Fascinated, I began to collect fossils immediately – a raw amateur, overwhelmed with curiosity and furious with my own ignorance.'[3]

She was the first to give serious and sustained consideration to the Lightning Ridge fossils. Apart from a few that had made it into the collections of the Australian Museum in Sydney and the Natural History Museum in London, next to nothing was known and the material had been little studied. 'I had to teach myself what was a fossil and what wasn't', she tells me. 'That was a fairly lengthy process, but I learnt pretty quickly as soon as I took stuff to Ralph Molnar, a palaeontologist at the Queensland Museum, who was able to say, "That's not a fossil, and that is". He was a great inspiration. I took bits to Alex Ritchie at the Australian Museum too. He taught me to ignore everything that wasn't potch or opal.'

Other Lightning Ridge miners besides Elizabeth and Bob Smith had started to build their own collections of fossils. By 1984, brothers Dave and Alan Galman, who had amassed one of the largest collections, decided to sell. Some of the pieces were potch in blue, grey or amber; others were beautiful, sparkling gems. The pair wanted the material to remain in the country, so they arranged to show it to Ritchie of the Australian Museum. This led to Ritchie's first encounter, in a Sydney motel room, with what would become one of Australia's most significant fossils.

As the brothers laid out fossils on every available surface,

Ritchie came across one that made his jaw drop. 'The hair stood up on the back of his neck and his hands began to shake', Smith says. Just 3 centimetres long, the translucent fragment of jawbone didn't look like much, but at the time it was the oldest mammal fossil ever found in Australia. *Steropodon galmani*, as the species became known, was a primitive relative of the platypus and the discovery made the cover of the journal *Nature* in 1985.[4] That single fossil had pushed back the envelope of Australian mammal history by a factor of five: from 22 million to about 110 million years ago. At that time, *Steropodon*, a large platypus-like monotreme, lived on the same floodplains as Lightning Claw, near the Eromanga Sea.

Sometimes opalised fossils are translucent and reveal internal structures such as the channels that carried nerves and blood vessels, 'in absolutely astonishing detail' Smith tells me. Inside the *Steropodon* fossil, for example, experts spotted an unusually large mandibular canal, hinting that the creature had electro-sensory capabilities similar to today's platypus. Another species that has turned up a number of times is the bipedal herbivorous dinosaur *Fulgurotherium australe* ('lightning beast of the south'), which may have been a fleet-footed, herd-living species. One of the most eye-catching specimens is a thighbone preserved in brilliant cobalt-blue opal.

Not all opalised fossils come from 'The Ridge'. White Cliffs in New South Wales, Coober Pedy and Andamooka in South Australia, and a few other sites produce them too, but since most were under the Eromanga Sea itself, they preserve marine creatures. One famous example is a pliosaur dubbed 'Eric', a four-flippered marine reptile something like a modern seal, which was formally named *Umoonasaurus* in 2006. The 1.5-metre-long fossil is the most complete opalised skeleton known. A team led by Ritchie and palaeontologist Paul Willis spent 450 hours cleaning and preparing the fossil after it was acquired by the Australian Museum in 1986.

Sadly, the beauty and financial value of opalised fossils means many have been lost to science. Before the 1990s, many miners didn't understand the scientific value of the fossils, and countless specimens were broken up and cut into gems. 'The destruction and waste of these miraculous objects has been tragic', laments Smith. Even if the fossils do survive intact, they are highly sought after on the international market. Once specimens have vanished into private collections in Asia, Europe or the United States, their value to science is lost, and papers cannot be published on these fossils. 'There are laws against export of opalised fossils, but opalised fossils are usually small, easily ferreted away overseas into private collections ... Their intrinsic evolutionary data remains unappreciated, their mysteries remain unravelled and everyone is deprived of the chance to wonder, to be delighted and to learn.'

Tales of fossils that have gone missing are heartbreaking. There's the beautiful little perfect shark's tooth in sparkling green opal that left Australia with a Danish man and was subsequently advertised for sale online for something like AU$120 000, 'just the most astronomical amount of money', Smith says. Then there's the sauropod vertebra in blue opal, as big as a house brick. 'An extraordinary thing with solid blue opal swirled around in black. That went off to Tucson [Gem and Mineral Show in Arizona] for sale in about '93, never to be seen again.' There was an entire fish backbone from the Carters Rush or Grawin opal fields, she adds, 'a whole string of opal vertebrae two foot [60 centimetres] long'.

There's a story of a Ridge man who has part of an opalised dinosaur jaw with the teeth still attached that he always carries on him, his most prized possession. 'There's another piece of dinosaur limb bone in town, with multiple colour bars', Smith says. 'Probably about eight ounces [225 grams] of high-quality opal. That's an extraordinary piece. If it ever gets into a public collection it will be a good day.'

Other tales tell of remarkable specimens Smith has never seen herself, but she has no doubt there is some truth to some of them at least – opalised dinosaur bones as high as your knee that were broken up to make gems in the early days of the fossil rush at the

Coocoran Opal Field, 30 kilometres west of town, and articulated specimens with whole groups of bones preserved together in the skeleton. 'There must also be many that we never even hear about, that go overseas.'

One mining claim near the town's airstrip once had three-toed footprints in the roof, perhaps hundreds of them – the miners were looking up from underneath at an ancient river flat. Before experts had a chance to look at them, the roof of the mine fell in. Another mine had much bigger footprints, possibly from an ornithopod similar to *Muttaburrasaurus,* and these were lost when the mine became waterlogged.

'This is all in the last 30 years. You can just imagine what must have been lost in the 70 years of mining before that', Smith says. 'There were early accounts in the *Walgett Spectator* newspaper of something like a little cat skeleton in pink opal [likely a dinosaur] that came up, possibly from the Three Mile Gully.' Then there was miner Joe Wopenka, who in the 1950s found the articulated specimen of a huge animal in potch, very probably a dinosaur. Word of it was sent to the Australian Museum, which eventually sent a request for part of it to be sent by train to Sydney. Joe was mining by hand, however, and digging out the estimated 2 tonnes of material and getting it to the nearest rail station 80 kilometres away, over dirt tracks, would have been a huge job. Ten years later the opal field was open-cut and the fossil was lost.

In the 1990s, Smith and her husband Bob lived as caretakers at processing dams at the Coocoran, where the miners 'wash' opal-bearing rock and dirt by running it through 'agitators' – modified cement mixers, which are a Lightning Ridge specialty. There they had fossicking rights to tailing heaps left behind by the miners. The extraordinary fossil material they collected included a huge range of plant and animal remains, even mammal jaws.

In the early days, Smith would package up small bundles of fossils and send them to Molnar to identify. He recognised bird bones (at the time only the third record of birds of this age in

Australia) and a tooth of the 8-metre-long herbivorous ornithopod *Muttaburrasaurus* – an exciting find as it was the first record of this Queensland species in New South Wales. Nevertheless, it was difficult for her to convince the museum experts that Lightning Ridge was worth investigating in more detail, because the institutions had little money for long-term fieldwork. Henk Godthelp and Mike Archer of the University of New South Wales (UNSW), Sydney, however, were intensely interested in the Cretaceous mammal fossils of Lightning Ridge. In the 1990s, Henk led teams of palaeontology students to the opal fields a number of times to search for fossils underground, as well to hunt on the surface of the opal fields and through jars of rough opal, possibly containing undiscovered fossils, in the local gem shops. Henk also took claystone excavated direct from the mine face back to the laboratory in Sydney, searched through it meticulously, and found a wealth of tiny fossils, including fragments of theropod eggshell, a monotreme mammal tailbone, miniature crocodile teeth and minute bones from baby dinosaurs.

Out at the processing dams, Elizabeth continued to find fossils – pieces of turtles and crocodiles, freshwater snails and mussels, dinosaurs, plesiosaurs and more. 'It was just magic, and you never knew when you walked up the truck ramp what was there and what you were going to find', Elizabeth says. 'It was bliss on a stick. It really was. Every single day.'

Outback fossil dig

Smith and fellow palaeontologist Jenni Brammall, who first visited Lightning Ridge with the UNSW team, are the scientific brains behind the Australian Opal Centre (AOC), a museum at Lightning Ridge. The AOC has a major collection of opalised fossils, donated by miners and many others throughout Australia. The first time I spoke with them was in 2008, when I was writing a short story about opalised fossils for the science magazine *Cosmos*. I knew then that the specimens were unique, and I'd always wanted to cover them in *Australian Geographic* too, so when I received an invitation from Brammall to participate in a fossil dig in Lightning Ridge, I readily accepted. *Australian Geographic* now collaborates with the AOC on annual digs where scientists and volunteers hunt for fossils for the museum's collection.

So it was that on a stifling October afternoon in 2014, under the brilliant-blue big skies of the outback, that I arrived for a week of finding, sorting and identifying fossils at one of Australia's most productive and significant fossil sites. A fossil dig at Lightning Ridge is not what you would expect if you've been digging elsewhere. The miners themselves have done the heavy lifting, excavating the 110-million-year-old deposits from 10–20 metres below the ground, so there isn't much digging involved. Instead we pick through discarded mullock heaps on the opal fields, an activity that locals call 'specking' – this involves holding your face close to the ground and searching for any small flash of colour or unusual shape that suggests a fossil. We also sift through unwashed fossil-bearing claystones and tailings, donated by miners.

Hunting for fossils in any context involves patience, typically slow pacing up and down with eyes to the ground looking for any hint of the unexpected. Hunting for fossils in the mine tailings is done sitting at a table with a pile of rock in front of you, but it isn't an altogether different experience, and I find the simplicity and all-consuming nature of it curiously enthralling. On my second dig at the Ridge we also make moulds and plaster casts of the beautiful blue–grey bones of Lightning Claw – including the 15-centimetre-long, grappling-hook-like claw.

Most of the time, picking through the tailings is a hushed, meditative activity, people mostly concentrating with their heads down, but the quiet is intermittently broken by a clamour of excitement when somebody finds a great fossil and everybody crowds in for a closer look. Exciting finds made during the digs I've spent at Lightning Ridge include fish jaws, bits of turtle carapace and belly shell (plastron), sauropod teeth and ornithopod vertebrae. For me, some of the most arrestingly beautiful specimens are tiny pine cones and spiral freshwater snail shells. One of the most startling was a tiny, perfect, opalised theropod tooth with flashes of red and steak-knife-like serrations along its leading edge – as precise as though it had just yesterday slipped from the dinosaur's mouth. That one was found on the surface of the ground only a five-minute stroll from the AOC's main shed.

The large number of theropod teeth and bones at Lightning Ridge is one of the great mysteries of the AOC's collection. As carnivores are at the top of the food chain, in most dinosaur sites around the world their fossilised remains are greatly outnumbered by herbivore remains. In North American fossil assemblages, for example, plant eaters outnumber predators by about 10 to one, but in Lightning Ridge the theropod teeth occur just as frequently as teeth of ornithopods and sauropods. Could it be that the predators hunted and scavenged in the billabongs, frequently shedding teeth as they went? Only time and more research will answer that question.

Only three species of dinosaur have so far been described from Lightning Ridge, but there are numerous dinosaur specimens in the AOC's collection of many thousands of fossils that have yet to be studied and described. When Bell returned to Australia after a decade working on duck-billed dinosaurs with Phil Currie in Canada, he was excited to approach the AOC. 'Dinosaur palaeontology has limped along for the last 80–100 years in Australia, hindered by the fact their bones are very rare here', he says. 'When I returned in 2013, top of my list of places to visit was Lightning Ridge. I had no idea of the diversity or richness of the fossil material, but I knew it was unstudied. When the AOC opened up its vaults I was gobsmacked. I immediately saw I would be spending the rest of my career here.' Bell is hopeful that he will soon be describing as many as 10 new species of Australian dinosaurs based on Lightning Ridge specimens, and he argues that it is one of Australia's best preserved ancient ecosystems in terms of range and diversity of species. 'We find an incredible richness of plant material, which is evidence of a very rich and diverse environment that was heavily forested', says Brammall, who adds that the site is unusual for producing numerous plant and invertebrate remains in the same deposit as vertebrate fossils.

Hitting the mother lode

Another Lightning Ridge miner, Bob Foster, hit upon an incredible discovery in 1985, even though he had no idea of its significance at the time. For many years he'd been pulling small amounts of opal out of the white clay in his mine and occasionally fossils too, sometimes even large ones. But opal mining is a fickle business, and making ends meet can be difficult, so fossils with colour would often be cut into gems. Nobody realised back then that the fossils could be very valuable too.

'We didn't know what it was, all we were interested in was looking for a bit of colour in them. And what we used to do is smash them up. We smashed probably a trailer load', Foster recounted to ABC radio.[5] But one day he found something big and, his curiosity piqued, decided to hang on to it. There was enough to fill a couple of overnight bags and some other crates. Foster took a train to Sydney and arrived with his unusual find at the Australian Museum, where he dragged the bags up to the office of Ritchie. 'I said, "I've got some bags full of dinosaur bones". 'He must have been thinking, "Here's another one".'

But on examination, Ritchie realised the bones were those of a large herbivorous dinosaur and asked to hang on to them for further examination. Within a short while, he and fellow museum palaeontologist Robert Jones headed out to Lightning Ridge with a group of Australian Army reservists to help Foster excavate the fossil bed in his mine. It became clear that it was very difficult to find fossils by excavating them in the fashion of a more conventional dig. Nevertheless, they turned up dozens more bones and took out large claystone blocks with bones still embedded. Unfortunately, the expedition was also the birth of a myth in the community that if the government found out you had fossils in your mine, the army would be sent in to confiscate them, at gunpoint if necessary. 'This is still Ned Kelly country up here', Smith says. 'People really resent bureaucratic interference, and some don't understand that the government has no legal right to take opalised fossils from miners.' (Since then Smith and Brammall have worked hard to better inform locals, and now many miners understand the importance of the specimens and are donating them to the AOC under the Australian Government's Cultural Gifts Program, which rewards donors with a tax deduction.)

Fifteen years later, Foster's fossil – which we now understand is the most complete dinosaur from New South Wales – remained unstudied, and when Foster returned to Sydney for a visit he was perplexed to find it on display in an opal store. At that point, he says he 'got the shits with them down there' for not asking permission, and decided to reclaim his fossils and take them back to Lightning Ridge. After that, Foster lent the dinosaur to the AOC, which gave it pride of place in its public display, with a label acknowledging him as the owner. Still the fossils remained unstudied.

That was until Bell took up a position as a dinosaur scientist at UNE in Armidale, which at 500 kilometres away is one of the closest universities to Lightning Ridge. He was keen to see the largely unstudied dinosaur collection at the AOC, and was astounded when Brammall and Smith guided him over to what he describes as a 'beautiful skeleton'. Bell soon realised the remains of this *Muttaburrasaurus*-like herbivore represented one of the most complete dinosaur skeletons ever found in Australia. 'My jaw dropped immediately. I never realised there was such fantastic stuff right here in New South Wales', he says.

At that stage, the Foster family decided to donate the fossils to the AOC, clearing the way for formal scientific research. Bell has since taken the fossil-bearing claystone blocks back with him to

Armidale, where CT-scanning techniques have allowed his team to probe the inside of the rock virtually, without having to extract the bones from the block. A large vertebra, as well as a rib and a bone from the shin are visible from the outside – but there are other elements hidden inside too. Using three-dimensional digital reconstructions you can manipulate these bones on a computer, just as if you were holding them in your hands.[6] Most excitingly, one of the blocks contains part of the skull, which Bell says is the 'holy grail' when working on a new specimen. From the skull you can get some of the best information about what an animal looked like, how it lived and what it was related to.

'Now begins the long, slow process of formally describing this animal. In order to name a new species, a palaeontologist has to literally describe each of the bones in a skeleton and compare those bones with the bones of related species', Bell says. 'This is a long, drawn-out process and the publication process takes considerable time as well.' The dinosaur is yet to be officially described, but Bell says Foster is likely to feature in the name when it does. Asked about this on the radio, Foster said, with a nonchalance typical of Lightning Ridge miners: 'Makes you feel old don't it ... I suppose it's good if you're into that sort of stuff. It'll be a good talking point down the pub'.

Impossibly unlikely and impossibly beautiful

Lightning Claw, the Gondwanan megaraptorid we met at the start of this chapter, was found by miners in the mid-1990s at the Carters Rush Opal Field, 60 kilometres from Lightning Ridge. The fossil consists of a giant claw from the hand, as well as parts of the arm, hip and foot, pieces of rib and a variety of other fragments. Sadly, the original specimen may have been much more complete but was partially destroyed during excavation, as the miners didn't realise at first what they had found. Opal miners Rob and Debbie Brogan acquired the fossil and in 2005 donated it to the AOC.

Skilled work by local preparator Matthew Goodwin revealed new details, and Bell says that when he first laid eyes on it in 2013 he realised it was a really significant find. The shape of certain features immediately hinted that the remains were a megaraptorid dinosaur – and larger than anything found before in Australia. Lightning Claw had unusually massive claws, which Bell believes it could have used like grappling hooks to ensnare prey, given that megaraptorids lacked the huge jaws and massive bite force of theropods such as *T. rex*.

A number of questions remain unanswered about Lightning Ridge. How many species of dinosaur were there? Why the puzzling proportion of carnivorous dinosaurs? Why does the fossilisation process favour small objects? Why have no fossils from very small mammals been found? The biggest question of all is how warm-climate species, such as crocodiles, turtles and pond snails, were able to survive in an environment that was at a very high palaeo-latitude, possibly right on the edge of the Antarctic Circle. Carbon dioxide levels were high at this time – the Albian stage of the Early Cretaceous (100–113 million years ago) – and would have warmed the planet, allowing the tropics to expand north and south, but still there would have been long periods of complete winter darkness. It's hard to imagine how those reptiles, which rely on the sun for warmth, would have been able to survive, and yet it appears they did.

The AOC is in the process of developing a large new facility to house its national collection of opalised fossils and gems, and a research centre where scientists can work at answering some of these questions. 'These fossils represent an irreplaceable national resource', says Smith, 'yet until now they have never really been accessible to the Australian people or to scientists'.

'These are the most outrageous dinosaurs on the planet', says Brammall. 'One in a billion plants or animals leaves a trace of itself in a fossil. Then the formation of opal is incredibly rare; it happens in very few places on earth. And we have these two things coming together in these incredible objects. It is impossibly unlikely and they are impossibly beautiful, and they tell huge stories about the history of our continent.'

The opalised fossils of Australia are among the great treasures of the world, she adds. 'Every object in the collection is packed with stories, not just of the ancient life and the geology, but the history and the stories of the miners who discovered these things.'

Discoveries elsewhere in what was once Gondwana are treasures of a different kind, expanding our understanding of dinosaur evolution in surprising ways.

9
RECORD-BREAKING TITANS

PATAGONIA, ARGENTINA

The titanosaur sauropods of Late Cretaceous South America reached epic proportions, the largest weighing nearly as much as two jet airliners. A flood of new Patagonian fossils is helping solve the biggest dinosaur mysteries of all — how and why did they get so large?

Small geothermal pools bubble and simmer, green-tinged and caked around their edges with a scum of pink microbes and a yellow crust of salts and sulfur. Some of the larger ponds intermittently spew forth noisy geysers of scalding mineral-rich water and steam. No plants grow in these hot pools or the volcanic soils that surround them, but there is life here aside from the heat-adapted algae and bacteria. A short distance away from the water, the whip-like end of an enormously long tail flicks into view, then one massive foot, and then another, meet the ground with a reverberating thud, sending up clouds of dust. One of the largest land animals that ever lived is using her pillar-like hind legs to dig out a hollow in the earth where she will carefully lay her clutch of about 30 round eggs, each the size of a large grapefruit.

To witness this breathtaking scene, we have travelled back 101 million years to the forests and floodplains of Patagonia, lush

with ginkgo and monkey-puzzle trees and horsetail ferns, and then part of the southern supercontinent of Gondwana. This a very special time in earth's evolutionary history, for it is in this window of time around 80–100 million years ago that the true giants of the Mesozoic arrived. This newly found species of titanosaur, excavated in 2014, had a 2.4-metre femur or upper leg bone, much taller than the average man. Size estimates suggest it could have looked into a seventh-story window with ease, and that it weighed perhaps 70 tonnes and measured 38 metres from head to tail. That's roughly equivalent in weight to 23 SUVs, 1000 people or 10 *T. rexes*. This likely makes it larger than *Argentinosaurus*, the previous record-holder, but with only six of that species' bones discovered, it's difficult to know for sure. Until 2013, the largest of the titanosaurs (a group represented by about 70 species worldwide) were only known from a handful of bones, so many of our ideas about them were based on inference and guesswork. But the discovery of this new titanosaur and the description of a slightly smaller species, *Dreadnoughtus*, in 2014, has revolutionised our understanding of these fascinating dinosaurs.

Having excavated her hollow and deposited her eggs, the female uses her mouth to heap vegetation on top of them. Sauropods were unable to sit on their nests, and many of them used the heat from decomposing vegetation to incubate their eggs. But this species has an added trick in its arsenal, as this site is a huge titanosaur nesting ground, covering many square kilometres, and here hundreds of dinosaurs return to deposit eggs each year in the warm volcanic soils. In this respect, these titanosaurs are similar to crocodiles and modern megapode birds (such as Australia's scrub turkey and mallee fowl), which use rotting vegetation; and the Polynesian megapode or malau of Tonga, which lays eggs in volcanic ash to take advantage of the geothermal heat.

Once hatched from these 15-centimetre-wide eggs, thousands of sauropodlets will swarm across the landscape and run the gauntlet of waiting predators, including carnivorous dinosaurs *Giganotosaurus* and *Tyrannotitan*, and prehistoric crocodiles and snakes, all of which arrive each year to benefit from the remarkable bounty of hatchlings – much as predators such as dingoes, sharks and seabirds prey upon tiny turtles emerging from coastal nesting sites in Australia today.

A record-breaking discovery

In 2008, Argentinean farm labourer Aurelio Hernandez was on horseback, searching for lost sheep on a 12 800-hectare ranch known as La Flecha on a plateau in the heart of Patagonia. Here, 260 kilometres south-west of Trelew, the regional capital of the province of Chubut, is a harsh and unforgiving landscape of red rock and low vegetation. It is a place of stifling summer heat and bitter winter cold. Something caught Hernandez's eye – a piece of rock that looked out of place and that would turn out to be a 2.4-metre-long femur, the largest thigh bone anyone had ever set eyes upon. Suspecting it might be part of a dinosaur, Hernandez reported it to siblings Alba and Oscar Mayo, the owners of the property. After a few years they reported the find to fossil technician Pablo Puerta at the MEF in Trelew.

'Perhaps 10 times a year somebody comes to the museum saying they have a dinosaur egg or a bone, and I have to tell you that 90 per cent of the time, these fossils turn out to be just rocks', says MEF palaeontologist Diego Pol. 'But you have to go and check out every report you get. So we went there and what this guy had found was the tip of the femur, the hind-limb bone, sticking out of the ground. This was the first discovery of a unique and fantastic fossil site that we were extremely lucky to find in the middle of the Patagonian desert.'

The MEF team set up a camp at La Flecha and began to dig a quarry that got progressively larger as they found ever more bones. Most dinosaurs – such as *Argentinosaurus* – are represented by just a handful, so Pol and his team were staggered as they continued to find more and more bones. Split across seven expeditions, the excavation took 18 months, and required the creation of new dirt roads that would allow trucks to carry the bones away safely encased in plaster jackets. By the end of the excavation in 2014, they had found 213 bones, which came from six individuals and together covered about 70 per cent of the skeleton of the species (sadly, not the skull – they have only been found for about three of the 70 known titanosaurs).

The site was nothing short of a dinosaur cemetery, says Pol, and – even more fortuitously – layers of white sediment between the red rocks were volcanic ash laid down in successive eruptions, meaning they could date the deposits very accurately. Volcanic ash includes zircon crystals with small amounts of uranium; through radioactive decay at a measurable rate the uranium turns into lead. By measuring the ratio of uranium to lead, experts can obtain a very precise date for the when the eruptions occurred. Great dinosaur localities including layers of volcanic ash are very rare, so this was a remarkable piece of good luck and returned an age for the fossils of 101.6 million years.

After the initial find they had to wait anxiously for the entire remains to come back to the laboratory at Trelew and be prepared and cleaned before they knew if they had a new species or if the fossils were those of an exisiting titanosaur. 'After you've prepared all the fossils is the first time you have a really good idea of the anatomy of these animals. By comparing the anatomy with the known bones from *Argentinosaurus* and all known titanosaur species, then you can know for sure if you have a new species', Pol says. Unique features that defined the dinosaur as a new species included very tall neural spines – projections sticking up from the top of its vertebrae – and also the shape of the channel through which its spinal cord ran.

When I speak to him in 2016, Pol and his team are planning to give the species a name that honours the location in which it was found and the people who originally discovered it. This is the first fairly complete specimen of a giant titanosaur and the first chance to get a good impression of what they were like, says Pol. 'The most amazing thing for me was not only the size, but the number of remains in a single site. Now we have most of the skeleton, this is the first time we can think about answering questions that even a few years ago were impossible to answer … This is a unique opportunity to understand how these animals lived, moved, and grew to reach these gigantic sizes, so it's really, really exciting for us.'

Calculating the weight of sauropods is a controversial task, especially when different research groups are pushing to prove the new species they've just found is a record-breaker. As Pol and his team have such a complete composite skeleton from the six individuals they found, they were able to use a tried and tested method, which involves taking the circumference of both the femur and the humerus (upper arm bone), and extrapolating the total body mass using data from living animals as a guide. There is a good correlation between the weight of an animal and the size of the bones supporting that weight. Remarkably, they found that the specimens from La Flecha were not yet fully grown, as indicated by so-called 'sutures' within the vertebrae that hadn't fully closed up. This means the full adult size of this species was perhaps even a little larger than the 38 metres and 70 tonnes they estimated from the specimens.

Additional clues found at the site, and anatomical estimates made on the titanosaur skeletons back at the lab, revealed a wealth of other details to fill in the story. Teeth of the *T. rex*–sized predator *Tyrannotitan*, and bite marks on some of the sauropod bones, hint that the corpses were scavenged after these animals died. Each of the collections of bones were found in slightly different soil horizons too, suggesting these giants did not die at the same time, and may have ended their lives here thousands or more years apart. Furthermore, the attachment scars of the creature's giant muscles on its hind limbs suggest that the swing of its tail helped its back legs move efficiently, similar to how Komodo dragons get a push from the movement of their tails today.

The new titanosaur is the second of several relatively complete Argentinean titanosaurs discovered in recent years. In 2014, researchers led by Ken Lacovara, then at Drexel University in Philadelphia, announced a new 26-metre, 60-tonne species they named *Dreadnoughtus schrani*, in reference to the formidable battleships of the early 20th century.[1] This animal would be 'nearly impervious to attack' from predators, they wrote in a paper describing the find. The fossil was first found in 2005 in Patagonia's Santa Cruz Province, and initially Lacovara's team was disappointed by what looked like a badly broken-up set of remains – but after four field seasons and a lot of digging they'd found 145 bones from two individuals. Before this discovery, the most complete titanosaur was *Futalognkosaurus*, known from 27 per cent of its skeleton, but one of the *Dreadnoughtus* specimens had 45 per cent, with unique bones in the other specimen nudging up the total known a little higher.

The discovery of several substantially complete titanosaurs

in recent years is allowing us to build a comprehensive picture of these animals for the first time, and is answering fundamental questions such as how their muscles were arranged and how they moved.

Hitting on supersized potential

The massive size of sauropods is not their only peculiar characteristic. Research by palaeontologists Michael Taylor, at the University of Bristol in the United Kingdom, and Mathew Wedel, at the Western University of Health Sciences in California, have shown that the incredibly long necks of these dinosaurs is a trait that has never been seen in any other group of land animals. Their research, published in 2013,[2] showed that no other group has achieved a neck longer than 2.5 metres (this is not counting marine reptiles, such as plesiosaurs, which achieved neck lengths of 7 metres or more, with an astounding 75 neck vertebrae).

The pair believe that the reason behind this is a confluence of factors in the evolutionary ancestors of the sauropods. The first dinosaurs appeared around 250 million years ago in the Triassic and were small, generalist, bipedal creatures. They evolved into a variety of major forms, including the ornithischian herbivores, the sauropod group and the theropods. The progenitors of the sauropods themselves were mostly medium-sized creatures that were starting to adopt the shape of their larger descendants, but they still spent some of their time walking on their hind legs, perhaps to reach high vegetation. They would have looked something like the 9-metre-long prosauropod *Lufengosaurus* from Yunnan in China or *Glacialisaurus* from Antarctica. Some of these animals soon hit on a less energetic method of reaching foliage on the crowns of araucaria pines (such as monkey-puzzle trees) which could be 40 metres high. By 215 million years ago, early sauropods such as *Antetonitrus* from South Africa had arrived at the form familiar to most schoolchildren – a big dinosaur with a long neck and tail, and a small head, walking on four pillar-like legs. *Antetonitrus* was 8–10 metres long and weighed about 2 tonnes, but by the Late Cretaceous, the largest Patagonian sauropods were more than 30 times as heavy.

Taylor and Wedel believe that the first clue to how this became possible lies in the fact that theropods and sauropods (but not ornithischians such as *Triceratops*, with big heads on short, stout necks) retained a trait of their ancestors – tearing off chunks of food and swallowing it without chewing. Chewing necessitates large teeth and jaws, and significant musculature for grinding, which results in a relatively large and heavy head. Instead, sauropods gulped plant matter down whole and relied on bacteria in their fermentation-chamber intestines to digest it. To this end, most sauropods had very small heads equipped with tiny brains that weighed just 25–50 kilograms. They had peg- or pencil-like teeth that were ideally suited to rapidly stripping fresh vegetation from branches, and their long necks meant they could move from tree to tree over wide areas without having to move their bodies.

The largest living land mammals, such as elephants and giraffes, are limited in neck length by the weight of their heads. In giraffes this is about 2.4 metres, a length never beaten even in the largest prehistoric mammals, such as mighty indricotheres that weighed up to 15 tonnes. There are several further reasons why this might have been the case. Mammals have a set number of seven neck vertebrae, which remains true from giraffes and baleen whales to bats and humans. Sauropods didn't have this limitation, and their neck vertebrae ranged from 12 to 17 in number.

Sauropods also share a number of the features of their anatomy with birds, which allowed their necks to grow ever longer. For a start, their bones had a honeycomb-like structure, filled with cavities, and were 50 per cent or more air by volume. CT scans have shown that the Chubut titanosaur's 15 huge neck vertebrae are about 80 per cent air. In the theropod dinosaurs this ancestral trait would eventually allow some of them to take to the skies as birds, as it lightened their load for flight, but in the sauropods it allowed their necks to become very long and their bodies to grow enormous, giving them much lighter frames than expected.

The second trait they share with birds is a very efficient set of lungs. Birds have a one-way system of air sacs beyond their lungs that passes throughout their bodies, infiltrating their skeleton and limbs, and drawing oxygen to their extremities. If sauropods had this set-up too, they could have moved oxygen much more efficiently throughout their system, making a large size much easier to achieve. In addition, the total volume of lungs and air sacs is much larger than simple mammalian lungs, which means the volume of air inhaled can be larger. In mammals the length of the neck and trachea is limited by the size of the lungs, because there must be enough fresh air drawn in with each breath to fill the lungs. If the volume of air sacs in sauropods was larger than their lungs, it would have allowed them to have much longer necks than mammals. As the necks of sauropods grew, the length of their robust tails had to grow as well, to act as counterbalances and keep their centre of gravity in the region of their midriffs. These were animals so huge they had to sleep standing, for fear they might never rise again from a resting posture.

The driving force for ever more growth

Assuming sauropods did have a unique evolutionary potential for long necks and enormous bodies, what might have been the driving force for them to grow to ever larger sizes? Attaining huge size entails complex engineering problems and is incredibly expensive in terms of energy, so there has to have been a very good reason why evolution took them in that direction.[3]

One theory is that animals that don't chew their food but rely on microbes to ferment and digest it for them benefit from having a longer gut, which allows for more processing time. Some experts believe the huge size created a feedback loop requiring sauropods to eat ever more vegetation and requiring ever more gut length to process it. This resulted in them growing in size until they reached something near the limit of what is biologically possible for a land animal. 'These animals must have had the physiological toolkit to reach those sizes. So they not only needed the food resources to grow and maintain a body like that, but also, from a physiological point of view, their digestive organs needed to be highly efficient to be able to capture all the nutrients they needed', says Luis Chiappe of the Dinosaur Institute at the Natural History Museum of Los Angeles County in California.

Another idea is that the huge size of sauropods was driven by a predator–prey arms race with the enormous carnivorous dinosaurs that harassed them. 'As some of the carnivores are getting bigger, herbivores are also keeping up with that', explains Chiappe. 'You have some really colossal carnivores living at the same time – things like *Mapusaurus* and *Giganotosaurus* – that were likely preying on these animals. These were the apex predators, and you can speculate that maybe they were preying on the young and the weak or hunting in packs. To some extent, the evolution of titanosaur sizes is a response to the predation pressure. It's a mechanism for defence.'

British researchers Graeme Ruxton of Glasgow University and David Wilkinson of Liverpool John Moores University argue that the long neck was an adaptation that allowed sauropods to sweep their heads back and forth, and collect as much food as possible without having to move their immense bodies around. They liken this engineering solution to a cylinder vacuum cleaner. Long necks were selected for in evolution as they allowed for feeding over wide areas from a standing position, thus reducing the 'energetically expensive' movement of the entire dinosaur. 'We draw analogy here with the cylinder vacuum cleaners that were commonplace in households in industrialised countries from the 1950s through to the 1970s', they wrote in *Biology Letters* in 2011.[4]

Because the machinery required to create the suction was large and heavy, the main body of the vacuum cleaner was positioned by the user in a central location within a room, and the user then moved a light head-part at the end of a long tube across the surrounding carpet ... the long neck of the sauropods might have been an adaptation to allow less movement of the exceptionally heavy body of these animals.

Using *Brachiosaurus* as an example, their calculations suggested that it would have used about 80 per cent less energy foraging for plant matter with a 9-metre-long neck than with one that only stretched to 6 metres. 'While vacuum cleaners have evolved over the past half century into smaller, lighter, more agile beasts, sauropods were wiped off the face of the Earth by the chance arrival of an asteroid 65m[illion] years ago', a journalist in the United Kingdom's *Guardian* newspaper noted humorously.[5]

Although we can't be sure that the new titanosaur at La Flecha in Patagonia was a species that migrated annually to nesting grounds, we have very good evidence of sauropod migrations from other fossil sites, such as those in North America. In recent years, clever studies have started to tell us more and about dinosaur behaviour – using clues that are not immediately obvious from looking at fossilised bones. One such study examined the chemical make-up of layers of enamel in the teeth of an American sauropod called *Camarasaurus*. It tells us that these 18-tonne, 20-metre-long creatures made annual migrations of up to 300 kilometres – something like the migrating wildebeest on the Serengeti today.[6]

Ratios of several isotopes (slight structural variants) of oxygen vary in water from different kinds of environment, such as lakes and rivers, and these end up in the teeth of dinosaurs via the water they drink. Through analysing sauropod teeth from Wyoming and Utah with layers of enamel laid down over many months, researchers led by geologist Henry Fricke at Colorado College in Colorado Springs showed the sauropods had been drinking water in several different environments. The research suggests that, during hot and dry conditions, when water sources dried up on their favoured floodplain habitats, these animals would travel hundreds of kilometres into upland areas with more regular rainfall. This migration allowed them to continue to consume the huge quantities of food their massive bulks required. 'They must have had the resources year round that were required to not just grow, but also sustain those sizes', says Chiappe. This makes annual migrations a very likely scenario for sauropods, much as huge herds of American buffalo once had to range far and wide across the prairies to find enough grass to feed themselves.

One remaining puzzle is why the titanosaurs of Patagonia were so much bigger than their relatives in other parts of the world. The predators here, such as *Giganotosaurus*, were even bigger than *T. rex* – but that can't be the whole answer, as predators, such as *Carcharodontosaurus* in North Africa, reached similar sizes too, but the sauropods there, such as *Paralititan*, only reached lengths of about 30 metres. 'I'm not entirely sure what the environmental factor was that allowed the animals to get so big in Patagonia', Chiappe says. 'I don't think that anybody has an answer.'

There is a chance that, as other continents that were once part of Gondwana are more extensively searched for fossils, huge titanosaurs will turn up in those places too. 'The titanosaurs you find in Argentina were there when Gondwana was to some extent fractured, but not fully. Given the land connections that existed for some time between Africa, South America, Australia and Antarctica, I would imagine it would make sense to find them', Chiappe says.

Eggs as far as the eye can see

There are a number of possible explanations as to why Pol's team found six different individuals of the new titanosaur that had died at La Flecha at intervals covering perhaps many millennia – something was regularly drawing these animals here and perhaps causing them to die here too. It could have been that there was an ephemeral billabong, and the animals that remained in dry seasons were trapped and died whenever the water ran out. There's another more outlandish proposal – but we know it's plausible based on fossil finds elsewhere – that the site was a titanosaur nesting ground to which animals returned annually in the breeding season, much as noisy colonies of seabirds gather annually on offshore islands today. The volcanic ash in the quarry shows that the region was geologically active, and we have evidence from fossilised nesting sites in Argentina and South Korea that geothermal heat was a tool sauropods sometimes used to incubate eggs.

Palaeontologists Lucas Fiorelli and Gerald Grellet-Tinner have been studying a Cretaceous-era hydrothermal site at Sanagasta in La Rioja Province, Argentina, where sauropods nested repetitively and purposely. 'The evidence gathered at Sanagasta documents that a group of neosauropods must have used soil thermoradiance and moisture from a hydrothermal origin to incubate their exceptionally large eggs', the pair said in a 2010 paper detailing the find.[7] Fiorelli and Grellet-Tinner believe the site would have been something like the modern Norris Geyser Basin of California's Yellowstone National Park, and the dinosaurs would have laid their eggs within metres of heat-releasing features such as hot springs, geysers and mudpots.

The MEF team found no evidence of nests and eggs alongside their giant dinosaurs, but another Patagonian site several hundred kilometres to the north-west, in the province of Neuquén, is home to the largest dinosaur nesting ground ever discovered, and when Luis Chiappe found it by chance in November 1997 it forever changed the direction of his career. Auca Mahuevo, as the scientists dubbed it, is on the flanks of the volcano Auca Mahuida, and dates to 83–79 million years ago ('Mahuevo' is a contraction meaning 'more eggs' in Spanish).

Chiappe had been searching for unexplored fossil sites that would include the remains of early birds from the Late Cretaceous period. He knew from geological maps that this region had exposed strata of the right age, but when he and his colleagues began to walk across the cracked mudflats below banded ridges of red rock, they soon realised they had discovered something else entirely. Everywhere around them were grey fragments of rounded rock with a distinctive surface texture. These were dinosaur eggs, and they were so abundant the crew couldn't walk in some spots without treading on them. They were elated and amazed, for they had discovered, quite by chance, the kind of untouched and wildly fossil-rich site that most palaeontologists can only dream about.

'It's an amazing place. For me it was the most amazing project I have ever had in my life', Chiappe says. 'Imagine a place where you walk for miles and miles and miles and you can follow the same egg horizons in the rocks. There are uninterrupted fields of nests … You see these clutches of eggs that are separated by a couple of metres, but close to one another and dozens and dozens of them. It is just incredible.' More than 500 eggs were found at Auca Mahuevo. Many of the eggs retain portions of sauropod embryos inside, with tony fossilised bones and skulls, and in some instances even fragments of mineralised skin. 'You can still see the patterns of scutes that formed the external skin of the animal', says Chiappe. 'It's really a unique place.'

The dinosaur eggs here were laid by hundreds of titanosaurs in an old riverbed, which was flooded once more in a catastrophic event, smothering the nests in mud and preserving them. Details of nesting behaviour were preserved too, such as the fact the nests were dug as holes in the dirt and then covered by the parents with mounds of rotting vegetation, much as megapode birds do today.[8] The nests here each contained 15–40 eggs.

'We didn't find any early bird fossils, but instead we found this incredible site. It's the role that serendipity plays in science', says Chiappe. 'You often go to a place looking for something and find something quite different, and that may determine a turning point in your career. I found this site and spent five or six years working on it and publishing a couple of dozen papers.'

South America's palaeontological potential

Along with other MEF palaeontologists, such as Pablo Puerta, Ignacio Escapa and José Luis Carballido, Diego Pol has spent the last decade patiently pacing through the dry, scrubby canyons and bluffs of the Chubut River valley, looking for telltale hints of fossils and new sites to quarry. Aside from occasional encounters with Patagonian wildlife, livestock and ranchers on horseback, they have the landscape pretty much to themselves. Many of the fossils, such as the new titanosaur and *Argentinosaurus*, come from the Late Cretaceous, but there are older Mid-Jurassic riches too, windows in time not well known elsewhere in the world, which offer further possibilities for groundbreaking dinosaur discoveries.

Dinosaur science in Argentina got underway under José Bonaparte in the 1970s and 1980s, but has exploded in the last 15 years, as a result of a more favourable political climate for research, and the appearance of papers – Chiappe's on Auca Mahuevo, for example – in major international journals, such as *Nature* and *Science*. Important fossil discoveries have brought fame and prestige to Argentina, as they have to China, resulting in increased government funding for research. Both Chiappe and noted palaeontologist Rodolfo Coria, who led the discovery of *Argentinosaurus* and *Giganotosaurus*, trained under Bonaparte at the Argentine Museum of Natural Science in Buenos Aires. Bonaparte, now retired, is among the most prolific dinosaur hunters of all time, and is credited with driving the birth of homegrown palaeontology in Argentina. Regional museums displaying local dinosaur treasures, such as MEF in Trelew and the Carmen Funes Municipal Museum in Plaza Huincul, are now thriving.

'Argentina, unlike any other South American country other than Brazil, has a palaeontological school', says Chiappe. 'These countries that have a long history of professional palaeontologists – as opposed to Peru, Bolivia, Ecuador or Chile, where there are just a handful of palaeontologists working – Argentina has an army of them.'

Patagonia has been very rich because of the desert terrain – dry, arid, minimal vegetation cover, not a lot of soil over the rocks – and also because lots of dinosaur-era Mesozoic rocks are exposed. The finds to date have been remarkable, but there is huge potential for further discoveries, argues Chiappe, as the majority of Patagonia remains very little explored. Much of the fossil hunting that has been done in this vast region of more than 1 million square kilometres is around roads, and determined by access, but there are huge tracts of inaccessible land that have been effectively off limits. 'I think there are going to be a tremendous amount of dinosaur discoveries coming out from that part of the world, and that's also going to be the case for the rest of South America', Chiappe adds. 'Currently we know very little about South American dinosaurs that are not from Patagonia.'

There are scant dinosaur discoveries from Brazil, Venezuela and Chile, but dense vegetation cover across much of the continent makes fossil hunting challenging. Some exciting recent finds hint at what to expect in the future, including: *Chilesaurus diegosuarezi*,[9] an oddball herbivorous theropod related to the tyrannosaurs that was described in 2015; the first dinosaur from Colombia, a brachiosaur called *Padillasaurus leivaensis*;[10] and Venezuela's first two dinosaurs – *Laquintasaura venezuelae*, a dog-sized herbivorous ornithischian, and *Tachiraptor admirabilis*,[11] a 1.5-metre-long carnivore – both of which date to 200 million years ago in the very early Jurassic and were described in 2014. There was also a spectacular fossil of an early snake with legs, which caused a storm of controversy when it was described in 2015, because it had been illegally exported from Brazil and ended up in a German museum.

But as more roads are cut, development picks up pace and expeditions are planned to arid areas with exposed rock that have never been thoroughly explored, such as the Bolivian Plateau, there are very likely to be significant discoveries, Chiappe says. 'The Andean nations – Bolivia, Peru, Ecuador, Chile – all have great potential, but very minimal work has been done. There are very few palaeontologists that work in those countries and not a lot of exploration at all.'

Palaeontologist Peter Dodson called sauropods 'nature's grandest extravagances', and it seems a fitting title for these implausible giants. South America has been a treasure trove of epically proportioned titanosaurs, and opened our eyes to the marvels of engineering possible in the natural world. Still more unexpected Gondwanan wonders, perhaps from other parts of the dinosaur family, are sure to come to light in the coming years.

Another part of Gondwana that has been yielding fascinating fossils in recent years is Madagascar, and though the dinosaurs here – including a series of titanosaurs – are close relatives of those in South America, they are telling a quite different story.

10

SOUTHERN KILLERS SET ADRIFT

MAHAJANGA BASIN, MADAGASCAR

Madagascar is renowned for its unique modern fauna, including lemurs and chameleons, but fossil-hunting expeditions over the past 20 years have revealed that Cretaceous-era Madagascar was also home to a rag-tag menagerie of peculiar carnivores, from goofy predators to horned cannibals.

Two feisty little dinosaurs decked out in blue-and-purple iridescent feathers are squabbling amid scrubby vegetation in the dry bed of a seasonal river. These males leap at one another, wings flapping, claws outstretched, hissing and nipping with jaws full of tiny, sharp teeth. Sporting vibrant, breeding-season finery, these magpie-sized carnivores are surging with hormones and will soon vie for the attention of a mate. Though they are close relatives of birds, these theropods do not have full flight feathers and cannot take to the air; instead, their plumage is primarily for display. Breeding season is a perilous time in the parched furnace of Late Cretaceous Madagascar. The climate here is similar to the outback of Australia or the deserts of Namibia 70 million years in the future – and this island is a place of long crippling droughts followed by brief flooding rains and subsequent boom times.

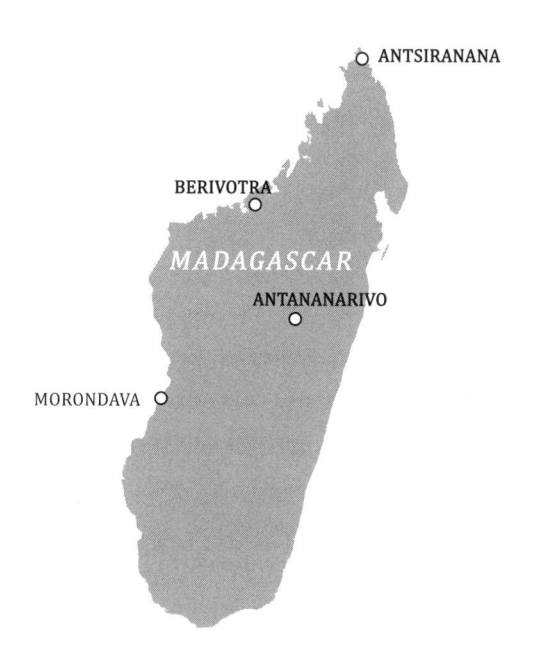

Distracted by their tussle, these little dromaeosaurs have failed to notice a 2-metre-long predator with short, downy plumage in black-and-brown stripes that is pressed close to the ground and creeping towards them with a flick of its long tail. This bipedal theropod is the kind of dinosaur you couldn't mistake for any other: *Masiakasaurus*[1] had teeth that stuck forwards and outwards from its lower and upper jaws, giving a goofy edge to its terrifying profile. When the conditions are good, this snaggle-toothed creature catches fish, but not today. It picks up speed, its dusty coat bathed in gold by the late-afternoon sun. By the time the *Rahonavis*[2] have noticed its approach, it is too late. The predator has leapt into the air and landed on one of the glossy, feathered creatures, clamping it between the interlocking teeth of its upper and lower jaws, and sending up a cloud of feathers. The lucky escapee darts off across the sand dunes, leaving a trail of three-toed footprints.

The plumage colours here are perhaps fanciful inventions, as are the behaviours, but many of these other details we know to be true from the fossils of these species. *Masiakasaurus* – revealed in a 2001 *Nature* paper titled 'A bizarre predatory dinosaur from the Late Cretaceous of Madagascar'[3] – was one of the smallest of the abelisaurids, a group of curious predators that flourished on the southern Gondwanan landmasses of Africa, South America, India and Madagascar. The northern continents were linked as the supercontinent of Laurasia, and there the tyrannosaurs – from *Daspletosaurus* and *Yutyrannus* to *T. rex* – ruled supreme. But the tyrannosaurs were never able to reach the southern landmasses, and here different breeds of carnivore spread and diversified. The abelisaurs were among the apex predators of the Southern Hemisphere, and on the island of Madagascar a variety of them – including the much bigger *Majungasaurus* and a new species, *Dahalokely*, described in 2014 – reigned. Other strange creatures that lived alongside the dinosaurs, and which we'll get to in more detail later, included a vegetarian crocodile and *Beelzebufo*, the horned 'devil frog' the size of a beach ball.

Seven-metre-long *Majungasaurus* in particular was a strange beast. It had something like the generalised body shape of a theropod dinosaur – standing on its hind limbs with a bipedal stance and small arms – but its snout was blunter and broader than that of a *T. rex* or an *Allosaurus*, and its face was gnarled and bumpy, with strange knobs, lumps and pits all over its skull. On top of the head was a small domed horn, and its arms were even tinier than those of *T. rex* – reduced to useless little appendages with fingers but no elbow or wrist joint. It's hard to imagine what purpose these could have served, and they may have been vestigial limbs that evolution was yet to eradicate.

Most intriguing of all is a 2003 finding that *Majungasaurus* cannibalised members of its own species, perhaps in response to the incredibly harsh environment during times of drought. Though cannibalism has been suggested in dinosaurs such as *T. rex*, it cannot be conclusively demonstrated that the bite marks on those bones are from the same species. With *Majungasaurus*, however, the evidence is definitive, making it the only proven dinosaur cannibal.

'It was a rabid meat eater anxious to rip into the flesh of whatever it found', co-author Kristina Curry Rogers, based at Macalester College in St Paul, Minnesota, said at the time of the discovery.[4]

The first humans to lay eyes on any hint of this strange creature would arrive 70 million years later in 1895, at the vanguard of an invading European army; and the next to give any detailed consideration to these prehistoric clues would not arrive until another century after that.

An experiment in evolution set adrift

In 1895, France invaded Madagascar, and among its soldiers were officers tasked with documenting the natural history of the land they were soon to acquire as a colony. The earliest record we have of dinosaur fossils in Madagascar comes from one of these French officers, who found some teeth, vertebrae and a claw at Berivotra (pronounced *ber-eeft*) on the north coast of Madagascar, about 400 kilometres north-west of the capital Antananarivo. The army officer sent them to Charles Depéret, a palaeontologist and geologist at the University of Lyon, who had spent 10 years of his career up to 1888 as a military doctor. Depéret described the carnivorous abelisaur *Majungasaurus crenatissimus* from these scrappy remains in 1896, although at the time he called it *Megalosaurus crenatissimus*. It was among the earliest dinosaurs discovered in the Southern Hemisphere (the first, South Africa's *Massospondylus*, had been named by Richard Owen in 1854). Little in the way of further remains would be discovered in this part of Madagascar for nearly a century.

When David Krause, a professor at Stony Brook University (SBU) in New York State, set out in 1993 to find fossils here once more, he needed some idea of where to focus his efforts, and so, after a little research, turned up Depéret's brief paper describing the finds at the Mahajanga Basin of Berivotra. Employing a similar method of retracing the footsteps of historic fossil collectors to that employed by the teams that had rediscovered *Spinosaurus* and *Deinocheirus* in recent years, he decided this documented fossil locality would be as good a place as any to start.

But, like Luis Chiappe at Auca Maheuvo, and so many other palaeontologists who chanced upon something unexpected, Krause did not set out to find dinosaurs. He was hoping to find mammals that were the ancestors to Madagascar's lemurs, tenrecs and other modern species, and answer questions about how the 587 000-square-kilometre island has come to have its largely unique modern fauna and flora. Madagascar is separated from Africa today by the swift currents of the 400-kilometre-wide Mozambique Channel, and about 90 per cent of its plants and animals are endemic.

Throughout much of the era of the dinosaurs, Madagascar was part of Gondwana, but during the Mid-Cretaceous a chunk of the southern supercontinent, made up of it and present-day India, began to drift north. By about 88 million years ago, Madagascar was isolated as an island, and it took with it a unique cargo of species that would evolve largely isolated from outside interference. The same thing happened with Australia, which has been an island for nearly 100 million years, and accounts for the fact that most of its mammals are marsupials and monotremes descended from ancestral branches of the mammal tree long extinct in most of the rest of the world.

'Today Madagascar is an incredibly rich area for biodiversity', says palaeontologist Andy Farke at the Raymond M Alf Museum in Claremont, California. 'You have groups of animals like lemurs and chameleons, weird carnivores and all different sorts of things that are very separated evolutionarily from their closest relatives in other parts of the world. The overarching questions is how did Madagascar's current set of plants and animals get to be there. That's what drives my fieldwork there, and intrigues other people who are working there too.'

It's not clear how they arrived – perhaps on rafts of vegetation – but somehow early primates made it on to the island and evolved into the array of about 100 species of lemur that we see today. Other species, such as shrew-like tenrecs and relatives of the mongoose, made it too, and they evolved into animals that filled the niches of hedgehogs, moles, otters and predatory cats – none of which had ever made it to Madagascar. One of the species I find most remarkable is the elusive fossa,[5] a relative of the mongoose that evolved into the island's largest predator and looks like a small cougar. There's also the enigmatic, 3-metre-tall elephant bird, which survived until the 1600s and is thought to have had the largest egg of any bird, larger even than those of sauropod dinosaurs.

We know little about the early evolution of these groups of species, or how they arrived, and this is what Krause was hoping to uncover. But this is not what he found at all.

A magical moment of scientific discovery

In his youth, David Krause assumed he would grow up to be a rancher. He had spent his childhood in the 1950s on a remote farm in the Canadian province of Alberta, where he says he was educated in a single-room schoolhouse and travelled around in a horse and buggy. He says he didn't even know what the word 'fossil' meant until a grain elevator operator from the nearby town of Hilda turned up one Sunday and asked Krause's father if he could go fossil hunting on their land. His father agreed and allowed the man to take Krause along on his hunt for bones along the South Saskatchewan River. 'I was surprised that my father allowed me to take a day off from work and do such a frivolous thing', Krause said in an interview.[6] 'I found a tail vertebra of a hadrosaur that day and must admit that it intrigued me to no end. It opened up a new world to me.'

Though he was fascinated, the possibility of becoming a palaeontologist didn't occur to him until his first year at the University of Manitoba, when a professor, Richard Fox, took him on as a field hand. 'I was immediately hooked', Krause says. 'The opportunity to live and work in the great outdoors coupled with the thrill of discovering animals long extinct seemed like an unbeatable combination. It still does today, and I'm as pumped about it now as I was then.'

The rest, as they say, is history. Following those early years, he worked on important fossil digs in the Bighorn Basin of Wyoming and the Crazy Mountains Basin of Montana, before arriving as a professor at SBU. He raised enough money for a foreign expedition, and in 1993 Krause and a small team of palaeontologists and students arrived in Antananarivo, where they had no idea what to expect.

They soon discovered it wasn't possible to turn up at the local rental car outlet and hire a few 4WD vehicles the way you could in the United States. The learning curve in this cash-strapped developing nation (which today has a population of 22 million) was steep. After a week, Krause had managed to purchase three vehicles of varying quality, all of which dated to the 1960s. One of them had been serving as a chicken coop, he says, until the owners found out they had a buyer and promptly refitted it with a motor and wheels.

After a 500-kilometre journey on dirt roads, they arrived at Berivotra in the middle of the night, before setting up camp and collapsing exhausted. Brimming with excitement for the finds that lay ahead, they awoke early the next morning and decided to set off for a promising outcrop of light-coloured sandstone just a short distance away. To Krause's disbelief, one of his graduate students found a fossilised mammal tooth within 20 minutes of them arriving and casting around on the ground. 'It was one of those magical moments of scientific discovery', Krause said later on.[7] 'We were jumping up and down screaming. I was yelling in French and I don't even speak French.'

The joy was short-lived, however – they would spend another six weeks there and during that time didn't find another single fragment of a mammal. But they did find plenty of other things – what Krause would describe as a 'mother lode' of Cretaceous fossils.

This was the start of a project that has since seen them return to these sites for field seasons nearly every year. In those early years of what has now been a two-decade project, they found hundreds of *Majungasaurus* teeth and pieces of bones, Krause says, but finding a skull became their obsession. In 1996, their hard work was rewarded with a great specimen; this also revealed that an earlier piece of skull they'd thought was from a dome-headed relative of the *Pachycephalosaurus* was in fact the base of the domed horn from a *Majungasaurus*.

Palaeontologist Peter Dodson, who was on that trip, says that local Malagasy colleague Florent Ravoavy took them to a site where an earlier French expedition had uncovered a number of dinosaur tail vertebrae. 'A bench on the hillside was literally strewn with bone, and fairly cried out in flashing neon lights "Dig here!"' Dodson wrote in *American Paleontologist*.[8] 'Here we dug, and here we encountered one of the most memorable specimens of any paleontological career: the complete disarticulated skull of an abelisaurid theropod, complete with menacing teeth.'

Majungasaurus is now one of the best understood Southern Hemisphere carnivores. The Mahajanga Basin Project – led by Krause at SBU and his colleagues at the Université d'Antananarivo – has uncovered two nearly complete skeletons, in addition to numerous other bones. The preservation of many fossils here is very good, which can be attributed to the fact that the flooding rains in the highlands at the start of the wet season would cause great debris-filled mudflows with milkshake-like consistency to wash down dry river channels like the one described at the beginning of our story.

Beginning to the south in the highlands and flowing out into the Mozambique Channel, these would smother and sometimes entomb the carcasses of animals that lay in their path – and even occasionally unlucky animals that were still alive. In some cases, the preservation of the fossils is so good that non-skeletal elements such as the keratin in claws (those of *Rahonavis*, for example) and ossified tendons are preserved too.

One discovery during the 1998 field season was the weird vegetarian crocodile mentioned earlier, *Simosuchus*. Krause's Malagasy graduate student Louis Laurent Randriamiaramanana unearthed a skeleton showing it had been a spaniel-sized, pug-nosed creature with relatively long legs and a squished-up tail. Rather than the sharp conical teeth of modern crocodiles, which are suited to snatching fish, it had leaf-shaped teeth, similar to those iguanas use today to shred vegetation. *Simosuchus* was also covered in osteoderms – bony plates in its skin – that gave it a covering of armour like an armadillo; this was no doubt useful for a creature that would have made a tasty snack for a *Majungasaurus* or the giant crocodiles also present.

These environments must have been dangerous, because *Simosuchus* wasn't the only armoured animal. A 2014 study revealed that *Beelzebufo*, the 'devil frog', which had previously been described in 2008, was also covered in bony plates. This beach-ball-sized amphibian may also have had small horns, giving it an appearance quite unlike any living frog. Intriguingly, its living relatives, sometimes called 'Pac-Man frogs', are fierce ambush predators found only in South America, hinting at the ancient link between these different regions of Gondwana. In the fossils of the dinosaurs, birds, crocodiles and mammals they find, the scientists working in Madagascar repeatedly find close links with South American species, says Curry Rogers. 'When we found out that that some of its relatives even have little horns on their heads, the "devil frog from hell" seemed an even more appropriate name', Krause told *National Geographic*.[9] In South America, these frogs will grab and gulp down pretty much anything that walks past. In prehistoric Madagascar, could they have fed on baby dinosaurs? It seems possible, but experts are yet to find conclusive evidence.

Another discovery from the late 1990s was the snaggle-toothed dinosaur *Masiakasaurus knopfleri*. The *Masiaka-* part of the genus name comes from the Malagasy word for 'vicious' – thus 'vicious reptile' – and also honours Mark Knopfler, lead singer and guitarist of Dire Straits. This was a wry nod to the fact his music had helped the scientists through the monotony of slowly chipping bones from the rock over a number of field seasons. Even palaeontologists have a sense of humour, it seems.

(Naming fossil creatures after rock stars isn't uncommon, according to palaeo writer Brian Switek.[10] There's also the 19-million-year-old, big-lipped mammal *Jaggermeryx*, honouring Mick Jagger of the Rolling Stones; the hoofed mammal *Gagadon minimonstrom* or 'Lady Gaga's little monster'; the 100-million-year-old hermit crab *Mesoparapylocheles michaeljacksoni*; and *Barbaturex morrisoni*, the 50-million-year-old iguana named for the Doors frontman Jim Morrison. When you're in the field you have a lot of time on your hands to come up with inventive names.)

After the initial description in 2001, a much more complete picture of *Masiakasaurus* was revealed in a 2011 paper[11] that examined different elements of fossils found in 30 localities, together building up a picture of two-thirds of the skeleton. The species probably used its weird front teeth to grab prey and its hind teeth to chew it up, a little like goofy-toothed South American shrew opossums do today with insects.

The mischievous giant lizard from Madagascar

Fossil sites are often in remote badlands and deserts, far from places of human habitation. But not at Berivotra. Here the fossil digs are often right within the village itself. 'The field area is really beautiful', says Curry Rogers. 'We're in the centre of a small village and the workings of it are all around us. We'll often be digging with kids watching over us, and there's traffic, and people walking by with their goats or carrying charcoal on their heads to sell. At the time when we first began there were huge language barriers, as the local people spoke a dialect of Malagasy, and not many spoke French, which was the national language … It's gotten easier over time as we've got more experienced.'

In the early years of the project, after noticing that some of the local children were following them around the field sites, Krause and his colleagues realised there was no school for them to attend – the community couldn't afford to build or staff one. To remedy this and give something back, Krause founded the Madagascar Ankizy Fund in 1998. *Ankizy* means 'children' in Malagasy, and the organisation's mission is to provide education and healthcare to kids in remote areas. Since building and opening its first school in Berivotra in 2001, the fund has now built another four schools in other villages, in addition to digging wells to provide clean water, renovating an orphanage, and flying in doctors and dentists.

Around Berivotra are dense grasslands in which surface fossils can be difficult to spot, but a chance bushfire burning off the vegetation of one outcrop led to the discovery of what is now one of the world's best known sauropods, and also the first ever titanosaur skull. It was during the excavation of this species – later named *Rapetosaurus*[12] by Curry Rogers and her supervisor, Catherine Forster, then of the State University of New York (SUNY), that they found something else entirely unexpected – the remains of the small feathered dinosaur *Rahonavis*.

Curry Rogers has been hooked on long-necked dinosaurs since she was a child, when she first encountered these 'incredible monsters' hatching from eggs on the illustrated cover of a magazine. She couldn't understand how anything so big could walk on land, and she wanted to know how such tiny creatures could grow so huge. She has since spent much of her professional career trying to answer this excellent question.

In 1995, when Curry Rogers was a graduate student at SUNY and arrived in Madagascar with Krause and Forster, the dinosaur bones they found most frequently were from sauropods. Few others on the expedition seemed much interested in them, she says, and most were focused on meat eaters, crocodiles and birds. Curry Rogers, however, thought sauropods were 'really cool', and , alongside Forster, set about digging up the titanosaur on the blackened hillside stripped of vegetation. The dinosaur they found was a largely complete juvenile about 40 per cent of adult size. 'The specimen was recovered from a single locality that had within it the remains of *Rahonavis*, just tucked in alongside some of the bones of these bigger sauropods, and also crocodilians', she says. '*Masiakasaurus* is also very well known from this locality.'

The scientists believe that these animals probably died at a time of environmental stress during a drought and had gathered at an ephemeral waterhole. When the rains return here, they mobilise fine-grained sediments that roll all of these remains together into bone beds. That explains why they often find different kinds of animals in close proximity. 'This is how you end up getting the tiny bones of a [bird-like dinosaur] alongside a giant sauropod', Curry Rogers explains. 'I have never worked in any place in my life where the bones are so beautifully preserved and so diverse ... you usually wouldn't find these delicate small things and these really big bones that are not so delicate together in the same spot, but in Madagascar you do.'

As they reported in a 2001 *Nature* paper,[13] they found both the skull of an adult titanosaur and the skull and skeleton of the juvenile. The discovery of the skull was very exciting – titanosaur fossils are notorious for lacking their heads, a frustrating phenomenon for palaeontologists (see the previous chapter). Not only did this *Rapetosaurus* fossil reveal that titanosaur nostrils were between their eyes and on top of their skulls rather than at the end of their snout, but it also suggested that the group was more closely related to sauropods such as *Brachiosaurus* rather than *Diplodocus*.

Curry Rogers and Forster asked their Malagasy colleagues at the Université d'Antananarivo to help them come up with a name for the species that had some local significance. They suggested several mythical creatures from folklore. One was a 'bad giant' called Darfefe that killed people and ate children. That one sounded like a pet poodle and was a little on the infamous side for a sauropod. Instead, they settled on Rapeto. 'It was just a mischievous giant who would just do things like step on a house by accident, but it wasn't doing anything bad on purpose. And so we named the dinosaur *Rapetosaurus*, which means "the mischievous giant lizard from Madagascar" for its first name', Curry Rogers said in an interview. The species name, *krausei*, honoured Krause, 'a really crafty thing to do just before your PhD defence', she adds.[14]

'*Rapetosaurus* gave us a kind of key for unlocking the relationships of all these other fragmentary specimens of titanosaur from all over the world', she says. That find was more than 15 years ago, and today hundreds of bones of *Rapetosaurus* have been found, along with as many as 10 really nicely preserved skeletons with many of their bones. Other species of sauropod may be more complete, but are known only from a single skeleton. There's also the developmental perspective. 'We know something about what it looked like when it was very small, and we also know something about what it looked like when it was very big, and every stage in between, and that's pretty remarkable and doesn't happen very often', Curry Rogers says.

The work on *Rapetosaurus* has continued in recent years, and Curry Rogers and her team are currently studying the growth rates of these dinosaurs and the skeletons of several tiny babies they recently found. Another research finding, published in *Nature Communications* in 2011, showed that the osteoderms found in the skin of these animals may have helped them survive the harsh conditions of Late Cretaceous Madagascar.

The function of titanosaur osteoderms had presented academics with a puzzle, because they are shaped differently from those of animals that used them for defence, such as crocodiles. In one of the adult *Rapetosaurus* fossils they found a huge osteoderm – the biggest ever discovered in any animal – about 57 centimetres long and 27 centimetres broad at its widest point. Curry Rogers says that finding that it was hollow was 'completely unexpected. There are no modern osteoderms in any vertebrate that are hollow in this manner'.

It turned out that the animals were using these bones as a calcium reserve they could call upon when they needed to construct the shells of their eggs. 'We think these osteoderms are hollowing out as the animal gets older, essentially as a way of having a mobile calcium reserve in the skin ... In the very stressful environment in which it lived, it would have needed a way of mobilising calcium to shell its eggs. Having these mobile minerals in the skin is a really handy way of getting around stealing minerals from your bones in your skeleton', Curry Rogers says.

Another area of research involves baby dinosaurs. In 2016 the scientists published a paper showing what life was like for hatchling sauropods, offering a never-seen-before picture of this vulnerable time in these dinosaurs' lives.[15] The largely complete skeleton of a young *Rapetosaurus*, which was between 39 and 77 days old when it died from starvation, shows that these animals grew very rapidly (from the equivalent of a chihuahua to a great Dane in about six weeks), and had already assumed their adult body shape at a very young age. Curry Rogers believes that this means they were self-sufficient from birth and could walk and feed themselves almost immediately after hatching. 'This baby's limbs at birth were built for its later adult mass; as an infant, however, it weighed just a fraction of its future size', she told reporters. 'This is our first opportunity to explore the life of a sauropod just after hatching.'[16]

Despite the huge number of finds so far, the scientists are confident there are still many new dinosaurs to be discovered in Madagascar. Among all the remains, for example, they have found a series of sauropod vertebrae that don't fit into what they know of the anatomy of *Rapetosaurus*. Curry Rogers believes there is at least one and maybe two or more new sauropods to be described – and that's just in the material they have already studied.

In 2014, Curry Rogers and her colleague Jeff Wilson described a new sauropod from a partial skull discovered by Andy Farke. They called it *Vahiny depereti*, which mean's 'Depéret's traveller' and honoured the French naturalist who'd named the first Madagascan dinosaur in 1896. It's possible some of the weird vertebrae are from the same species, but until palaeontologists find more of these kinds of remains in association with one another they can't know for sure. 'There are absolutely a lot of unanswered questions about the identity of at least one and maybe two or more sauropods from Madagascar', Curry Rogers says. 'Every time we go there we're finding tons of new fossils, and it takes a really long time to put all of those things together. The more work in the field, the more likely it is that we'll find something new.'

There are many specimens on museum shelves, such as at the Field Museum in Chicago, that have not yet been studied or even removed from their plaster jackets. There's also material from past years that has not even been shipped from Madagascar to the United States for preparation yet (the Mahajanga Basin Project eventually plans to return the fossils to Madagascar, once there is a museum or institution to take them). 'There's just a huge backlog of things we need to prepare, and sometimes we find really interesting things when we look inside the plaster', Curry Rogers says.

She is one of few female professional palaeontologists in what remains a very much male-dominated field. As of 2015, only about 17 per cent of Paleontological Society members were women,[17] and the number of women working specifically on dinosaurs (such as Australia's Jenni Brammall and Elizabeth Smith, see chapter 8) is even fewer. Curry Rogers believes the reason is partly down to the difficulty of balancing family life with travelling for collecting trips in the field – though there are many female palaeontology students these days, a much smaller proportion of them end up as working palaeontologists. Her fieldwork often involves six-to-eight-week trips to regions such as Zimbabwe, Patagonia, Montana and Madagascar, and taking her young daughter with her wasn't always an option, so she took turns with her geologist husband when it came to going away for field seasons.

There is no doubt some chauvinism in the field, and perhaps resistance to women becoming palaeontologists from some, but

Curry Rogers believes things are changing. She points out that, as of 2016, three of the past five presidents of the US Society of Vertebrate Paleontology had been women.

Opening up whole new horizons

One of the most recent dinosaur finds in Madagascar, published in 2014, was an early relative of *Majungasaurus* called *Dahalokely* – and that discovery was made at a new fossil site in a region of the island previously unexplored by fossil hunters.[18]

Andy Farke started working in Madagascar in 2003 when he was an SBU graduate student on Krause's team. The fossils from the Mahajanga Basin all date from around 70 million years ago, when Madagascar had already been long isolated as an island. But before this, Madagascar had been joined to India; the two were set adrift together in the Indian Ocean around 88–100 million years ago, when the shard that became Madagascar finally splintered off and India continued its slow sail north towards Asia. Farke wondered what the animals of Madagascar were like when it was first isolated and how they related to animals in other parts of the world. He wondered whether they were most closely related to known species from Africa or South America.

To help answer these questions, he and his colleague Joe Sertich of the Denver Museum of Nature and Science in Colorado had to find Cretaceous-era rocks even older than those at Mahajanga, from around the time Madagascar was separating from India. 'I wanted to see if I could find dinosaurs and other animals that might tell us something about the early stages of Madagascar right after it became completely isolated', Farke says.

Farke and Sertich started by examining geological maps of Madagascar created in the 1950s and 1960s. They identified a town called Antsiranana (formerly Diego-Suarez) – a 700-kilometres drive north of Berivotra on the island's far northern tip – that had rocks of the right age, about 90 million years old. A quick survey of the area with Google Earth also suggested there were favourable outcrops. Along with graduate student Liva Ratsimbaholison, from the Université d'Antananarivo, they set out to explore the site in summer 2007.

Just a few kilometres from the beach, the site is not what you typically expect of a palaeontological dig, says Farke. 'A lot of the image of fieldwork is that you're off in a desert or badlands, or something like that, with no vegetation, but at this site there are giant baobab trees everywhere and chameleons and cool wildlife. And if you walk up the hill from where we were excavating you could see the ocean. We were eating seafood pretty regularly.'

They didn't find anywhere near the volume of fossils regularly turned up at Berivotra, but they did eventually find the ribs and distinctive vertebrae of a new carnivorous abelisaur. 'Just a few days before we planned to depart the area, Joe Sertich found some bone poking out of a cliff. It looked to be a vertebra of some sort and a little digging revealed more bones', Farke said.[19] 'The hollow spaces in the bones inside made it clear they were some sort of dinosaur. This was a pretty happy moment, because we knew that dinosaur vertebrae can be exceptionally useful for identifying species.'

They didn't have time to collect everything, as the bones were encased in hard rock. Sertich and Ratsimbaholison had to return to finish excavating it in 2010. Following this, once they'd cleaned and studied the bones back in the lab at SBU, they could tell that they had once belonged to a medium-sized carnivore about 4 metres long. At the time of their publication, they were unable to tell which abelisaurids it was most closely related to, but Farke says they've since determined it is similar to later species from both Madagascar and India. They believe it is perhaps ancestral to the abelisaurs from both places – species such as Madagascar's *Majungasaurus* and its Indian contemporary *Rajasaurus* from the Narmada Valley. This would make sense, as the animal dates from around the time the landmasses separated.

They called the species *Dahalokely tokana*. In Malagasy *dahalo* is a 'thief or cattle rustler', while *kely* means 'little'. The species name, *tokana*, means lonely, and 'this dinosaur would indeed have been lonely, way out there in the middle of the Indian Ocean', Farke adds. The discovery 'reinforces the importance of exploring new areas around the world where undiscovered dinosaur species are still waiting', says Sertich, who has continued the hunt for new fossil-bearing localities in Madagascar.

He had heard rumours that there might be dinosaur fossils north of the city of Morondava and several hundred kilometres south of the Mahajanga finds, on the island's mid-west coast. Following the excavation in 2010, Sertich went to look for himself and was not disappointed. Almost immediately he found another titanosaurian sauropod, and says he had simply to peel back the grass to reveal a series of vertebrae and a hind limb. 'By the end of the first day we had a fully articulated tail exposed on the surface, so it was really easy', he says. 'Although it wasn't easy to get there – it was 20 kilometres from the nearest road and we had to use ox-driven carts to carry all of our gear. But in terms of finding a dinosaur on your first day, it was pretty spectacular.' He returned in 2012, and then 2015, and collected the rest of that skeleton, as well as another three partial skeletons of the same species.

The femur of the new species is 2 metres long, making this sauropod about twice the size of *Rapetosaurus* and roughly comparable to the 26-metre-long titanosaur *Paralititan* from Egypt (see chapter 1). His team is currently prepping the fossils back in Denver, and it will be some years before they are ready to describe the material. All up, across the three or four specimens, they have perhaps 65 per cent of the skeleton of the species. It dates from 85 million years ago, and, like *Dahalokely*, is one of the dinosaurs from shortly after Madagascar was separated as an island. That means this sauropod was alive about 15–20 million years before the species known from Mahajanga. 'A lot of the strange features of the younger dinosaurs probably resulted from isolation', he argues. 'Being alone on this relatively small subcontinent for 20 million years is what shaped that younger fauna into such a bizarre group of dinosaurs.'

The rocks of the Morondava basin have previously yielded a number of Jurassic dinosaurs, which were studied by the French from the 1950s onwards and are now held in collections in Paris, but no Cretaceous dinosaurs had previously been found there. The Cretaceous deposits are primarily near-shore marine environments, and the bloated corpses of these sauropods had probably floated out to sea before sinking and becoming fossilised. The sites are famous for yielding beautiful Cretaceous ammonites, of the kind you often see sliced in half and polished up in fossil shops.

Sertich has searched pretty much the entire west coast of the island for additional dinosaur-bearing localities, without any luck. He says, however, that there is now a lot more sauropod and theropod material coming from Mahajanga, but from a slightly lower layer, perhaps 2 million years older than the existing fossils. 'There's another 20 years of work in the classic localities', he adds. 'There are so many beautiful fossils'

Mammals at long last

It was the search for Mesozoic mammals that had first drawn Krause to Madagascar. But after the tooth his graduate student discovered in their first 20 minutes at the Berivotra field site back in 1993, it would be a long time before he found more interesting remains. In 2010 Krause finally made one of the most exciting mammal discoveries of his career – challenging what we thought we knew about the mammals that lived alongside dinosaurs south of the equator.

As with so many of the other discoveries, this groundhog-like creature – which Krause described as 'Punxsutawney Phil on steroids'[20] – was found pretty much by chance. In 2010, when Sertich was one of Krause's graduate students, he collected a huge block of sandstone he hoped was filled with fish fossils. Later, when the block was CT-scanned back at SBU, it became clear that there was also something much rarer and more valuable inside: the nearly complete skull of a prehistoric mammal known as a gondwanatherian. It was only the third time a Cretaceous-era mammal skull had been found in the Southern Hemisphere. This was the find of a lifetime for Krause, who called the 70-million-year-old animal *Vintana sertichi*, after the Malagasy for luck and also his student who had serendipitously brought its rocky tomb back to the laboratory. 'We know next to nothing about early mammalian evolution on the southern continents', Krause told reporters. 'No palaeontologist could have come close to predicting the odd mix of anatomical features that this cranium exhibits.'[21]

The previous few Southern Hemisphere mammal finds have added up to just a few teeth and scrappy jaw fragments. This nicely preserved skull was a much more useful fossil, and showed that the animal was relatively huge for a mammal of its time, weighing around 9 kilograms – two to three times the size of a modern groundhog. It would have had huge eyes and massive cheek muscles for chewing, the experts believe, and an unusual proportion of its brain was likely devoted to its sense of smell. But for now, many questions remain unanswered, and for those Krause is surely hoping the next find will be *Vintana*'s complete skeleton.

Krause may have been disappointed that instead of the many mammals he'd hoped for, he exposed a whole new world of Gondwanan dinosaurs, but on that point his co-workers don't agree. 'In my mind we got lucky. I'm a dinosaur palaeontologist, so I'm really happy we found dinosaurs', says Curry Rogers with a smile.

One piece of the Gondwanan puzzle that once linked Australia, South America and Madagascar is Antarctica – and despite the massively inhospitable nature of its frozen climate, palaeontologists have recently made many major dinosaur finds there.

11

POLAR PIONEERS AND THE FROZEN CRESTED LIZARD

TRANSANTARCTIC MOUNTAINS, ANTARCTICA

Incredible human tales of adventure and the fossils these heroic expeditions have uncovered have revealed that the world's ice-entombed southern landmass once teemed with prehistoric life.

On 17 January 1912, British Antarctic explorer Captain Robert Falcon Scott and his team finally made it to the South Pole. They had completed a gruelling trek over several months with only simple garments of wool, canvas and fur to protect them against temperatures that fell as low as minus 40 degrees Celsius. They arrived to find a Norwegian flag – planted 34 days earlier by Roald Amundsen, who had beaten them in the heroic race to our planet's southern extremity. 'The Pole. Yes, but under very different circumstances from those expected …' Scott wrote in his journal.[1] 'This is an awful place and terrible enough for us to have laboured to it without the reward of priority. Well, it is something to have got here.'

With heavy hearts, they began the 1300-kilometre return journey to their permanent base on the coast. As they dragged their sleds, Scott had already begun to fear that frostbite and the poor condition of his men might prevent them getting home. On Thursday 8 February, with a break in the weather, he ordered a half-day's rest before they continued down the Beardmore Glacier.

While on the other side of the world Ernst Stromer was uncovering dinosaurs in the Sahara, and Transylvanian palaeontologist Baron Franz Nopcsa was working as a spy for the Austro-Hungarian empire, Scott's team set up camp on a moraine under Mt Buckley. Despite their dejected and weak state, they prospected for and collected 16 kilograms of plant fossils. As Scott scribbled in his diary:

> I decided to camp and spend the rest of the day geologising. It has been extremely interesting. We found ourselves under perpendicular cliffs of Beacon sandstone, weathering rapidly and carrying veritable coal seams. From the last Wilson, with his sharp eyes, has picked several plant impressions, the last a piece of coal with beautifully traced leaves in layers, also some excellently preserved impressions of thick stems.[2]

Scott and doctor–scientist Edward A Wilson both knew that these fossils proved Antarctica had once been thick with vegetation, in stark contrast to the desolate world of ice and snow they saw around them. Scott had some sense of the importance of these fossils, given the team continued to carry them when they had jettisoned much of their gear, but they could not have known that these impressions of the prehistoric fern *Glossopteris* would ultimately provide the evidence for continental drift, confirming that Antarctica had once been joined to India, Australia and South Africa, where *Glossopteris* fossils had already been found.

They carried on down the glacier, hauling the fossils, meteorological logs, journals and photographic film on a treacherous route that crossed many crevasses. In the process, they lost one of their men, Edgar Evans, who fell and hit his head. By mid-March, with fuel and food supplies almost exhausted, they couldn't go on, and were trapped in their tent by the inclement conditions. 'We have struggled to the end and have nothing to regret', wrote Wilson to his wife, Oriana. Scott's last diary entry was on 29 March. A letter to his wife, Kathleen, regarding his two-year-old son, Peter, beseeched her: 'Make the boy interested in natural history if you can; it is better than games'.

Here, just 17 kilometres short of the next supply depot, the bodies of Scott, Wilson and naval lieutenant Birdie Bowers were found in their tent nine months later, where they remain today, frozen in their perpetually youthful state. What these adventurers and men of science had found changed our geological understanding of the earth. It was the most important outcome of the Terra Nova expedition, and was among the first of many fossil-hunting forays that have revealed the truth about Antarctica during the age of the dinosaurs.

Fiery reminders of Gondwana

Some of the earliest clues that the southern continents had once been connected came when botanist Joseph Banks sailed around the world on HMS *Endeavour* with Lieutenant James Cook in 1768–71. Banks found trees in Australia, New Zealand and South America that also became part of the evidence for continental drift and plate tectonics. The first piece of the puzzle came in January 1769, when Banks collected a specimen of Magellan's beech (*Nothofagus betuloides*) at Tierra del Fuego on the tip of Patagonia. He thought this 'southern beech' was related to the beeches and oaks of the northern continents, but we now know they are distinct family, possibly related to birches.

The crew of the *Endeavour* turned up more of these trees in New Zealand and Australia, where they arrived in 1770, becoming the first to sail up the eastern coastline. It wasn't until the 1830s, however, that another botanist, Joseph Hooker, began to puzzle over the similarities between species such as Magellan's beech, the myrtle beech from Australia and the silver beech from New Zealand. In Australia, one of the best known of this group of trees is the beautiful tanglefoot or deciduous beech (*Nothofagus gunnii*), also commonly known as fagus. It is Australia's only cold-climate deciduous tree, and its leaves turn brilliant, autumnal yellows and oranges in April and May each year before they drop. It grows in the highlands of the Tasmanian Wilderness World Heritage Area, where it is fiery reminder of the fact that Tasmania was once part of Gondwana.

We now know that there are about 35 southern beeches across South America, New Guinea, New Caledonia, New Britain, New Zealand and Australia. Southern beeches are also known as fossils from Antarctica – and, incredibly, may still have grown there in ice-free remnant patches just 3–5 million years ago. Of course, the reason all these trees are similar is that they share an ancestor that once grew across Gondwana, dominating its temperate forests.

The first southern beeches evolved more than 100 million years ago, and they are among the oldest lineages of flowering plants.

The *Glossopteris* fossils collected by Scott are today found at the Natural History Museum in London and the Scott Polar Research Institute in Cambridge, United Kingdom. In 1914 they were studied by British palaeontologist AC Seward, who made the surprising discovery that they were Permian-era seed ferns related to fossil species found on continents that we now know were all part of Gondwana, and, earlier still, part of the globe-spanning supercontinent Pangaea.

The idea that the earth's continents moved was first developed in detail by German polar scientist Alfred Wegener in 1912, but it was not widely accepted for another half-century. Today, everything we understand about the movement and distribution of dinosaurs is illuminated by our understanding of the changing shape of the world during the Mesozoic Era. Scott's *Glossopteris* were among the first fossils from Antarctica, a continent that has since yielded up specimens dating from 540 million years ago in the Permian through to traces of whales, dolphins and plants from just 2–3 million years ago.

Over the 75 years following Scott's discoveries, many more fossils of plants, fish and other creatures were found, but the first Antarctic dinosaur was not discovered until 1986, when Argentinean geologists Eduardo Olivero and Roberto Scasso came across the remains of an armoured ankylosaur, *Antarctopelta*, off the Antarctic Peninsula that stretches towards Patagonia. Here in Late Cretaceous, near-shore marine deposits on James Ross, Seymour and Vega islands, scientists began to find bits of other plant-eating and carnivorous dinosaurs.

Dinosaur discoveries have continued on the islands of the Antarctic Peninsula, with new species described in 2013 and 2015 and others awaiting publication. But the most spectacular Antarctic dinosaur discovery was made just four years after *Antarctopelta* and 3000 kilometres inland from the peninsula, near the Beardmore Glacier ...

The frozen crested lizard

In late December 1990, at the height of the austral summer, Ohio State University geologist and polar research veteran David Elliot was climbing to the upper reaches of Mt Kirkpatrick to peruse 170-million-year-old volcanic rocks. Joining him in this endeavour, just a short way down from the 4528-metre summit, was fellow geologist Richard Hanson of Texas Christian University. Part of the Queen Alexandra Range of the Transantarctic Mountains, Mt Kirkpatrick is near the Beardmore Glacier, where a US National Science Foundation (NSF) contingent of scientists was camped for its 1990–91 Antarctic expedition.

On a slope of scree 4000 metres above sea level, Elliot had spotted the telltale shape and texture of fossil bone. Realising he'd found a large animal, he radioed back to camp to alert a team of palaeontologists that had been working nearby in the Gordon Valley. Among that team was Bill Hammer of Augustana College, Illinois, who had just completed a month of studying Triassic fossils.

Over the next few days, with time fast running out, they used a generator and a jackhammer to try to remove as much of the hard, grey rock as possible. In slabs of mudstone that had formed on a prehistoric river bottom, they revealed the front half of a nicely preserved dinosaur. They had to borrow a small helicopter from the Italian Terra Nova Bay Antarctic base to get it down the mountain, as the helicopters then in use at the US camp were too large to land safely on the peak. But Bill realised the significance of the find and the importance of collecting as much of it as possible – the fossil would revolutionise what we knew about the distribution of these reptiles around the world.

In the end, the job proved more challenging than anyone could have imagined, requiring two subsequent expeditions to Antarctica, the most recent in 2010–11, but the discovery was immense. Antarctica was the last of the world's continents to offer up dinosaur fossils; this was not only the world's southernmost dinosaur quarry, at just 640 kilometres from the South Pole, it was also the highest altitude from which a dinosaur had ever been excavated. Furthermore, it was one of only very few Early Jurassic dinosaur sites in the Southern Hemisphere, in addition to those in South Africa and Argentina.

News of the incredible discovery at this most improbable location soon spread around the world, the *New York Times* reporting in March 1991: 'American scientists have found dinosaur bones on a windswept mountain near the South Pole, a discovery experts said proved beyond question that the celebrated prehistoric creatures were a global phenomenon'.[3] Elliot, who was also the chief scientist of the NSF's Beardmore Glacier research program told the newspaper: 'I wasn't looking for dinosaurs … It took us a few minutes to realise what we were seeing. Finally, the penny dropped for both of us. We realised we had found a new vertebrate locality for Antarctica'.

The new species was a medium-sized carnivore 7–8 metres long, which Hammer named *Cryolophosaurus* in a 1994 *Science* paper.[4] They estimated that it dated from about 190 million years ago in the Early Jurassic. This 'frozen crested lizard' was named for the strange crest on its head, and is believed to be related to other early crested dinosaurs, including *Dilophosaurus* from Arizona, *Sinosaurus* from China and *Dracovenator* from South Africa. These animals appear to form an Early Jurassic radiation of medium-sized crested theropods, which are an early branch of a much larger group of theropods, including birds.

With 50–70 per cent of its skeleton preserved in that single fossil, *Cryolophosaurus* is one of the best understood Early Jurassic theropods. It helped fill gaps in our understanding of the distribution of species across the single supercontinent Pangaea, which at that time was yet to split into Laurasia in the north and Gondwana in the south. Remarkably, the Mt Kirkpatrick site, no bigger than a football field, yielded up a series of other dinosaurs over the course of the three expeditions, but nobody had any clue in 1990 that this would be the case. The first of these was right under the *Cryolophosaurus*, below another 30 centimetres of sediment in the same quarry, and was discovered by Hammer once he began to uncover the fossils in the rock back in the United States.

As Hammer studied the fossils in detail, further details came to light – such as the teeth of other small carnivorous theropods and pieces of a flying pterosaur and a mammal-like reptile mixed in with the remains. The rocks containing these fossils had been laid down 190 million years ago in a river meandering across a lowland coastal floodplain. It was forested and dense with plants such as ginkgoes, conifers, ferns and mosses. (Another fossil site near the Beardmore Glacier has a ghost forest of fossilised tree stumps sticking out of the ground, some of which are 50 centimetres tall and more than a metre across.)

The Mt Kirkpatrick site was further north than today during the Early Jurassic, at about 65 rather than 85 degrees South, but would still have been inside the Antarctic Circle. Despite this, global temperatures were higher, and the climate here was temperate, comparable perhaps to the US Pacific north-west or New Zealand today. Antarctica formed the southern part of Pangaea, and would still have been in complete darkness during winter. The dinosaurs that lived there must have had special adaptations to endure these conditions, perhaps including fat layers, insulating coverings of feathers and large, sensitive eyes that made use of low light conditions. On the flipside, the vegetation must have enjoyed luxuriant growth during several summer months of 24-hour sunlight.

In recent work, palaeontologists including Peter Makovicky at the Field Museum in Chicago, where the *Cryolophosaurus* fossil is held, used a CT scanner to examine the skull and create an endocast of the brain. From this we can see that in shape and size it was somewhere between what you might expect for a bird or crocodile of the same size, with some bird-like structural features. Cross-sections of this *Cryolophosaurus*'s bones have revealed growth lines suggesting it was 15–17 years old when it died.

The brain wasn't the species' only bird-like trait. Its crest would have been a prominent structure on top of its head, bigger in life with flesh on top of the bone, and it very likely was used for display. 'If we compare it to living birds, we can see that birds that have a bony crest on top of their head tend to have a lot of colour on the face, not necessarily on the crest itself', says Makovicky, hinting at how this carnivore might have looked. Australia's southern cassowary, for example, has a brown crest and bright red and blue on the skin of its face and neck.[5] 'It's highly likely that *Cryolophosaurus* would have had some kind of feather-like plumage', adds Makovicky, pointing out that many dinosaurs later on in the same tetanuran lineage had feathers, as do modern birds.

Yet more Transantarctic discoveries

Phil Currie is an Antarctic veteran, having joined three dinosaur-hunting expeditions there, most recently in 2014. Antarctica is 'probably one of the coolest places I've ever been, not just in the literal sense', he says. Antarctica is 'pretty wild', agrees Nathan Smith of the Dinosaur Institute at the Natural History Museum of Los Angeles County in California. 'It's one of the most beautiful places I've ever been on the planet. The first time you step off the plane you just kind of take in the landscape because it's something else. It just goes on and on and on forever on a clear day.'

The first trip to Antarctica for both Currie and Smith was when Hammer finally managed to return to Mt Kirkpatrick in December 2003, 13 years after the initial discovery of *Cryolophosaurus*. Getting into the interior of the continent – which is twice the size of Australia or the contiguous United States and covered with a kilometres-thick sheet of ice – is expensive and logistically difficult. Many years can pass between the large-scale NSF expeditions that allow palaeontologists to get there.

In many ways, Antarctica continues to be as dangerous a place to work today as it was for Scott and his party a century ago.

Polar pioneers and the frozen crested lizard

A young Argentinian geologist once became hopelessly disorientated in white-out conditions on James Ross Island after stepping just 5 metres from a tent to get cheese from a storage locker. He survived the night and the blizzard by building a snow shelter, and found his way back to camp when the weather cleared 36 hours later, but he was lucky. His colleagues at the camp had tried to find him and were convinced he was dead. 'You try and be as careful as possible, as everything you do there is liable to result in disaster', Phil says. For him, the most hair-raising moment of the 2010 field season was when the weather conditions at the quarry 4000 metres up on Mt Kirkpatrick changed without warning. The team radioed for the helicopter that had dropped them off to come back and collect them, knowing full well it might not be able to make it through the cloud and fog. They carry survival gear including tents for shelter in an emergency, but at 4000 metres altitude in Antarctica there's no guarantee of making it through the night.

Thankfully, the pilot could get through. 'The problem on Mt Kirkpatrick is that you're just up so high and the air is so thin that helicopters are really limited in what they can do', Currie recounts. 'The pilot decided we would have to leave our gear and he would take everyone off the mountain, but by then the fog had surrounded us. He ended up lifting the helicopter up just slightly, making his way to the edge and then basically dropping off the side of the mountain. We fell into the fog and lost sight of everything, and then after we'd dropped a certain distance the blades started catching [as the air got denser] and we could come up again. It was a pretty frightening.'

They usually have two helicopters on the NSF expeditions, so they have search-and-rescue capability if one of them goes down. But in 2003, a helicopter that had been setting up relay stations for communications on one of the mountains got caught in a downdraft that smashed it against the rock and forced the skids up through the fuselage. The pilot managed to fly out, but the helicopter was out of commission after that. 'When we're making our plans we put down a large number of days for research time, but in reality we may only get a third of that because of mechanical issues, bad weather and other problems', says Smith. 'Particularly up at the dinosaur site. You might have days where it's gorgeous everywhere else, but if the clouds move in we can't get up there.'

Out of 28 possible days up at the quarry, they ended up with six, but they made the most of them, excavating another third of the *Cryolophosaurus*. Time constraints are among many problems that often require you to think on your feet in Antarctica. Because of the cold, the dryness and the altitude, for example, it can be difficult to work with glue and plaster, which are prerequisites for stabilising fossils on digs just about everywhere else. You also can't camp overnight on Mt Kirkpatrick, or you risk developing pulmonary oedema and other symptoms of altitude sickness. Spending 12–14 hours working hard there can give you a pretty good headache, says Smith.

A common misconception is that palaeontologists have to cut through the ice to reach fossils, but they actually have to find areas that are ice-free, both in the Transantarctic Mountains and on the islands of the Antarctic Peninsula. In total, less than 5 per cent of Antarctica has bare rocks. 'As with other parts of the world, you have to go where the rocks of the right age are exposed, but we also need to find those that aren't covered by ice. In that way, you have some different but similar problems to expeditions anywhere in the world.'

Smith had the privilege of studying the second dinosaur that had been found underneath the *Cryolophosaurus*. They extracted more of this in the 2003–04 field season, including leg, foot and ankle bones. Working with Diego Pol of Argentina's MEF in 2007, Smith described it as a herbivorous prosauropod dinosaur that would have weighed around 5 tonnes and reached a length of 7 metres.[6] He named it *Glacialisaurus hammeri*, honouring Hammer, who had been his undergraduate mentor.

Prosauropod dinosaurs probably looked something like the ancestors of the giant sauropod dinosaurs, although they are not their direct ancestors as they overlap with them in time. Prosauropods had long necks, but were much smaller than many of the true sauropods and some still spent part of their time walking on their two hind legs rather than four column-like limbs. Other prosauropods include *Lufengosaurus* and *Massospondylus*, dinosaurs from the Early Jurassic of China and South Africa respectively. The finds in Antarctica 'are important because they help to establish that primitive sauropodomorph [sauropod and prosauropod] dinosaurs were more broadly distributed than previously thought, and that they coexisted with their cousins, the true sauropods', says Smith.

As the *Glacialisaurus* was found directly under the *Cryolophosaurus* remains, an early idea was that the carnivore was feeding on it and had choked to death on one of its bones, but more study suggested that the fossils were not deposited on the prehistoric riverbed at precisely the same time. Instead, they had perhaps both been separately washed to a bend in the river.

Despite only having six working days on Kirkpatrick in 2003–04, Hammer's crew had the incredible luck of finding another herbivorous dinosaur. While the palaeontologists busied themselves with the specimens in hand, mountaineer and safety guide Peter Braddock scoured an area about 30 metres above them in a casual search for fossils. 'Keep your eyes down; look for weird things in the rock', Hammer had told him. Braddock found something weird indeed – part of a pelvis, which would in life have been much bigger than the corresponding bones of *Cryolophosaurus* or *Glacialisaurus*. This and other fragments of this animal were shipped back to the United States for preparation and study. The species is yet to be described, but findings so far suggest it would have been a sauropod and perhaps one of the largest dinosaurs ever found in Antarctica.

In 2010–11 they made it back to the Mt Kirkpatrick quarry for a third time, excavating the remaining third of the *Cryolophosaurus* and more of the *Glacialisaurus*. And – quite remarkably – making yet another fossil discovery. 'On a little hillside right behind the quarry, there was a place where we would sit at times, and we realised that here there was in fact another small dinosaur', says Currie. 'It was very different to the *Cryolophosaurus* fossil, which has black bones in hard rock. This was pink-boned and was a very small animal. We had to beg, borrow and steal more helicopter time before the expedition ended, and three of us went up and got almost the whole thing out in a day, because it was small and the rock was soft.' Back in the United States, they realised there were several individuals mixed in together. These specimens, which are still being prepared and studied at the Field Museum, represent juvenile prosauropods that had thigh bones just 20 centimetres or so long and would have weighed only 20 kilograms when they were alive.

'The new finds are potential sauropodomorphs. One is a semi-articulated skeleton, which includes a really nice skull and most of the skeleton, which is fantastic. This will be the most complete dinosaur from Antarctica probably, or very close to it with *Cryolophosaurus*', says Smith. 'So there were two, maybe three, species of these long-necked sauropodomorphs running around, and from our preliminary analyses, they don't seem to be close relatives of one another. You see hints of that pattern in other parts of the globe, too – for example, in the Early Jurassic of South Africa, where there are a number of sauropodomorphs living in the same environments that are not each other's closest relatives.' All of these fossils are from a key time for understanding the early evolution of dinosaur groups, say the experts, and they show us that true sauropods coexisted for some time with more primitive prosauropod dinosaurs in the same environments.[7]

The fact that they had found in a very small area the remains of five or six individual dinosaurs suggests the incredible fossil richness of the layer of rocks called the Hanson Formation. 'There are Hanson Formation outcrops on all of the adjacent mountains',

Currie says. Aerial surveys had allowed them to identify at least five other peaks where they could find outcropping sediments of the right age, but they had no time to prospect them.

Despite the likely profusion of dinosaur fossils waiting to be discovered at the Mt Kirkpatrick site and on adjacent peaks, the palaeontologists may not be returning there any time in the near future, as the following big NSF expedition to the Transantarctic mountains (2017–18) has the Shackleton Glacier, further south, in its sights. The exposed rock outcrops there have been more famous for Early Triassic fossils, which predate the dinosaurs. Expeditions to study these rocks will focus on answering questions about the Permian–Triassic mass extinction event 252 million years ago, which killed about 90 per cent of all species. Smith says that palaeontologists will also prospect for Mid- and Late Triassic rocks (from 201–247 million years ago), so there is a chance of finding some of Antarctica's very earliest dinosaurs: 'The Late Triassic is right now a blank slate for Antarctica for vertebrates in general. That means that, for the most part, anything we find there would be new to science. That's fairly exciting'.

The NSF expeditions to the Antarctic interior are logistically massive operations that cost many tens of millions of dollars and include vast numbers of scientists such as glaciologists, atmospheric scientists, geologists and climatologists in addition to palaeobotanists and vertebrate palaeontologists. In 2010–11, for example, there were 17 teams of scientists all focused on answering different questions about Antarctica. 'It's a multimillion-dollar operation. Unless you've got ten expeditions that all want to go to the same area, they're probably not going to do it just for you', Currie explains. 'I'd love to go back to Kirkpatrick, but it seems unlikely. Unfortunately, the locations of the expeditions are not decided on what you as the individual want, but on how many different projects they can fit together in one package to set up one of these major camps with an airstrip, all of the accommodation, kitchens, workplaces and everything else.'

It can be difficult prospecting at new sites if you have no idea whether you'll find fossils, and the geological maps of the region aren't good – for obvious reasons. You have to have already done your homework and have some sense of guaranteed results, Smith says. 'But that's also why there's the potential for a lot of new, cool stuff to come out of Antarctica. It's so hard to get down there and so little time has been spent exploring it compared to other places, such as North America, where fossil sites have been worked for hundreds of years.'

In that sense, the dinosaur hunters in Antarctica today are doing what prolific palaeontologists ED Cope and OC Marsh achieved in the west of North America in the late 19th century – they are for the first time revealing the dinosaur fauna of an entire continent.

The Naze theropod

The austral summer of 2003–04 was a productive period for Antarctic dinosaur discoveries. Less than a week after the new herbivore finds in the mountains in late December 2003, 3000 kilometres across the continent on the Antarctic Peninsula's James Ross Island, the scant remains of another dinosaur were found, on what once was the bottom of a shallow ocean. This 70-million-year-old find was the continent's first Cretaceous carnivore represented by more than a single bone.

To reach the site where they found the fossil, palaeontologists led by Judd Case, of St Mary's College of California, had to trek 13 kilometres each day across treacherous ice floes. 'We don't get many opportunities to go to Antarctica, and there is a short weather window of opportunity each time', Case said following the discovery.[8] But Antarctica 'consistently turns up new surprises ... I have no doubt that there is a tremendous fossil record buried under the ice sheet'.

Case, co-worker James Martin of the South Dakota School of Mines and Technology, and their team originally set out to test a theory about the migration of marsupials by looking for fossils on nearby Vega Island. They hoped to answer how early marsupials from the part of Pangaea that became North America eventually ended up in Australia – and were looking for evidence that Antarctica had acted as a migration route. Harsh weather, however, trapped the team's boat in ice, and they were unable to pursue their original goal.

Disheartened, they stopped off on James Ross Island instead. The rocks there are made up of sediments laid down in an ancient sea, so they didn't expect to find many land animals. Nevertheless, alongside fossil clams, ammonites and other sea life, more exotic fragments began to appear, including the legs, feet, jaws, teeth and bits of the skull of a theropod. It was likely the corpse had been a 'bloat and float' that died near the shore and was washed into the ocean before eventually settling on the bottom of the Cretaceous-era Weddell Sea.

Known as the Naze theropod,[9] after the island's Naze Peninsula where it was found, the dinosaur was first though by Case to have been a dromaeosaur similar to *Velociraptor*, but a 2015 reanalysis by Case and his student Ricardo Ely suggests the dinosaur was more loosely aligned to dromaeosaurs and perhaps other theropods, such as the troodontids.[10] Another Antarctic Peninsula theropod, a large carnosaur, is known from the discovery of a single bone in the 1990s by Ralph Molnar.

Some features of the bones of the Naze theropod and others from the peninsula hint that these dinosaurs were relatively primitive for their time. Some believe the Cretaceous dinosaurs of Antarctica represent a kind of relict fauna, consisting of groups more commonly associated with earlier times elsewhere on earth. One theory to explain this is that the newly successful flowering plants – the angiosperms – were slower to colonise Antarctica than other continents, perhaps because it was cloaked in total darkness for so many months each year. In the Late Cretaceous, Antarctica was still smothered with the cycads, palms and ginkgoes that elsewhere dominated during the Jurassic Period, and this may explain why more ancient types of dinosaur persisted there.

Though these islands of the Antarctic Peninsula are still difficult to reach, palaeontologists get to them more frequently than they do the interior of the continent. As of 2015, Rodolfo Coria's team had prospected for and excavated fossils at the peninsula 12 years in a row, flying in from Argentina. NSF-funded US teams also spend frequent summers here if their supply ship makes it through the sea ice. Palaeontologist Joe Sertich of the Denver Museum of Nature and Science was one of a team that prospected for fossils there in the summer of 2011–12. 'What really struck me was how pleasant the weather was', he says. 'It wasn't really cold, like a lot of people think. In fact, it rarely got below freezing. But we did have several big storms come in that were extremely windy. I remember being knocked off my feet by the wind. And the wind also brings in these crazy big icebergs, so the shoreline outside of our camp was littered by them after the windstorms. That can be challenging because it basically traps you. You can't get a ship through those floating piles of ice.'

Sertich planned to return in 2012 and 2013, but those field seasons were cancelled because of ice. 'There's a lot of future potential here and many possible sites, but it's a logistic nightmare getting in there, particularly further along the peninsula, where sea ice is an even bigger problem. Expeditions here are usually supported by ships rather than one of the main research bases. This usually involves getting to the exposures [fossil-bearing rocks] on small zodiac boats, and to do that it has to be completely free of sea ice.'

US expeditions to the peninsula are smaller than those to the interior, and can focus just on palaeontology and geology, though they are by no means cheap to put together. The NSF supplies the ship, helicopters, zodiacs and specially trained crew, and that alone may amount to US$60 000 a day.

An ankylosaur out of place

The first dinosaur to be discovered in Antarctica was the ankylosaur *Antarctopelta oliveroi*, found at a site in 1986 that also bears the remains of marine invertebrates and giant reptiles such as plesiosaurs. *Antarctopelta* would have been no more than 4 metres long, was covered in bony plates, scutes and spikes, and possibly had a club on the end of its tail. Armoured dinosaurs seem to have a habit of turning up mixed with sea life in near-shore marine deposits – as with fossils of the ankylosaurs *Scelidosaurus* in south-western England and *Kunbarrasaurus* in Australia – perhaps because they were common in coastal environments or because their physiology somehow made it more likely they would float and be carried out to sea.

In 2014, Currie and his wife, palaeontologist Eva Koppelhus, were out with Rodolfo Coria's team on James Ross Island not too far from where six dinosaur specimens had previously been found. They couldn't get to some of the sites because of sea ice, but they excavated more pieces of the original *Antarctopelta* specimen and found bits of mosasaurs and plesiosaurs, as well as shark teeth. The presence of an ankylosaur in Antarctica is interesting because they were rare in South America in the Cretaceous, although several have been found in Argentina and Australia. Ankylosaurs were much more common in Asia and North America.

Over the many years that Coria's team have worked on the peninsula, they have found a variety of other dinosaurs. As they reported in 2012, they found the vertebrae of a titanosaur sauropod on James Ross. 'We cannot do much with only a vertebra, so we don't know the genera or species', Ariana Paulina Carabajal, a palaeontologist then at the Carmen Funes Municipal Museum in Plaza Huincul told reporters.[11] 'But we know it's a titanosaur, it's a kind of sauropod that's very common in South America.'

Other recent Argentinian finds include *Morrosaurus antarcticus*, a medium-sized relative of the herbivorous ornithopod *Iguanodon*, known from a fragmentary hind limb that was found at the El Morro site on James Ross Island. The species was described in 2015 by a team led by Sebastián Rozadilla at the Museo Argentino de Ciencias Naturales in Buenos Aires, who noted that Cretaceous ornithopods found in Antarctica, Patagonia and Australia appear to be members of a similar Gondwanan fauna.[12] A related primitive ornithopod, *Trinisaura santamartaensis*, was found at the island's Santa Marta Cove and described by Coria and co-workers in 2013.[13] Coria named it after Trinidad Diaz, a female Argentinean palaeontologist and pioneer of the Antarctic Peninsula, hence the feminine suffix 'saura' rather than the masculine 'saurus' (it's one of few dinosaurs with a feminine name, including *Laquintasaura* from Venezuela and *Leaellynasaura*[14] from Australia).

Duck-billed hadrosaurs were another kind of dinosaur we know lived in Antarctica at this time, because a single distinctive tooth was found in 1998 on Vega Island. The only other Southern Hemisphere hadrosaur fossils come from South America. Experts believe hadrosaurs evolved in the Northern Hemisphere, then migrated somehow to South America, and from there to Antarctica. The chunks of Gondwana that became Antarctica and South America had already separated 70 million years ago, but there may still have been a chain of islands that allowed hadrosaurs to make the journey.

Fossilised remains of early members of modern groups of birds have also often been found, including *Vegavis*, from Vega Island, which was related to ducks and geese; and *Polarornis*, a diving bird something like a loon. These discoveries are evidence that modern bird lineages had already begun to diversify millions of years before the extinction that killed the non-avian dinosaurs. All up, with the remains of the ankylosaur and carnosaur, 'it's kind of a unique fauna for Gondwana at least', says Sertich. 'Some crossover of Laurasian taxa [such as duck-bills and ankylosaurs] with classic Gondwanan taxa, like the titanosaur sauropod.'

Antarctica was then sandwiched between Australia and South

America, and even after they began to separate it may have remained an important stepping stone for migrating groups of mammals and dinosaurs between these two regions. 'Antarctica was the gateway in and out of Australia at this time', says Steve Salisbury a palaeontologist at the University of Queensland in Brisbane. Salisbury was part of the NSF-funded Antarctic Peninsula Paleontology Project, or AP3, expedition that spent a month on Seymour and Vega islands in February–March 2016, collecting about a tonne of fossils, including some pieces of dinosaurs and numerous marine reptiles.

'Antarctica sat in the middle of southern supercontinent of Gondwana. It was connected to Australia and New Zealand in the east; South America in the west, via the Antarctic Peninsula; and centrally to Africa, India and Madagascar', Salisbury adds. 'So it's a really important place to test our ideas about the distribution and relationships of various plants and animals around at that time. The recent discovery of closely related megaratoran theropods in both South America and Australia has shown that some elements of Australia's dinosaurian fauna must have extended into Antarctica.'

Enormous Antarctic potential

Finding dinosaurs and other animals from the Late Cretaceous in Antarctica is important because it's one of very few Southern Hemisphere sites from the period shortly before the comet or asteroid impact that caused the extinction of much of earth's life 66 million years ago. Most data about the end-Cretaceous extinction comes from North America. 'The importance for us is filling in gaps about what we know was going on at high latitudes during the Mesozoic', says Smith. 'We talk about mass extinctions, such as at the end of the Permian or the Triassic, as global events, but they come with a huge geographic bias. Most of our fossil record is from areas that are more accessible and have been studied for a long period of time.'

Work in Antarctica so far has just scratched the surface, and – on Mt Kirkpatrick at least – numerous dinosaur remains have been found with relatively little survey effort. This means that future finds are likely to be very exciting, as long as the right rocks can be reached. Palaeontologists make it down there infrequently, but finds will filter out over the next few decades, and a number of new discoveries are already awaiting preparation and description. 'We need better data collection across Antarctica', Smith says. 'It's a huge place. We can then ask questions about whether there were different things going on with animals in different parts of the continent. We don't have a sample that allows us to ask those questions right now.'

Some of the plants and animal found there are similar to those in other parts of the world, such as South Africa, he says. 'We can start to see how they were differentiating in the Early and Middle Jurassic and can address how Antarctica started to become this really unique place, climatically, geographically and biologically.' Antarctic dinosaurs may have been different in ways that scientists don't yet understand. For example, some Australian dinosaurs from far southerly latitudes appear to have had large eyes and optic nerves – such as the small ornithopod *Leaellynasaura* – useful in a nocturnal habitat. Did some of them hibernate? Studies of growth lines in fossil bones from southern Australia so far seem to suggest these polar dinosaurs grew consistently throughout the year, but nobody knows for sure.[15]

One really key question for researchers is understanding precisely why and how Antarctica entered the deep freeze it is in today. At the end of the Cretaceous, although the continent was at roughly the same latitude as it is today, the conditions would have been temperate and seasonal – but not tropical, as people often imagine in the age of the dinosaurs. It is likely that the final separation of the elements of Gondwana – the decoupling of the Antarctic Peninsula from Patagonia and the separation of Australia sometime between 25 and 40 million years ago – that caused cool ocean currents to be diverted right around Antarctica rather than up the coast of South America into warmer latitudes. This cold current was effectively then isolated from the rest of the world, plunging the continent into deep-freeze conditions. A feedback loop smothered the continent with ice, reflecting the sun's energy away from it and back out into space.

Intriguingly, Antarctica may not have frozen over entirely until very recently. There are fossils of Antarctic beeches related to Australia's fagus that date from just 3–5 million years ago. Could it be that strange, cold-adapted Antarctic land mammals survived until then, too? Who knows what fascinating discoveries lie hidden beneath the thick cap of ice? I, for one, am hoping some of them will be revealed in the coming decades.

FUTURE POTENTIAL

Where to from here? Buckle up, because the new discoveries are going to come thick and fast...

What this world tour of dinosaur discovery has shown more than anything is that we never know where the next great find will turn up. We've covered plenty of weird dinosaurs and a lot of ground – from the Arctic to the Antarctic and Australia to Eastern Europe; and from implausible winged monsters and shaggy, pot-bellied enigmas to hump-backed filter-feeding curiosities and creatures decked out in an extravagance of feathers. But there's so much more I was unable to include. These are both new discoveries in regions with an existing history of palaeontological work and also new places with dinosaur-bearing deposits that have yet to be properly prospected.

Just a few of the new locations with enormous potential are Greenland, Myanmar, Uzbekistan, Kazakhstan, Turkmenistan, Tunisia, Malawi and Niger. For example, Michael Ryan at the Cleveland Museum of Natural History has found dinosaur footprints in Greenland and believes the island has great potential for further discovery if funding and local government support can be secured. That could give a whole new dinosaur fauna from a largely unexplored part of the world.

Other regions, such as Tunisia and Algeria, are likely to be as rich in fossils as neighbouring Morocco. Federico Fanti, of the University of Bologna in Italy, has been digging fossils in Tunisia, but political instability there and in Algeria means they are currently unsafe places for Western scientists to work. With wealth and development, many parts of the world currently inaccessible to foreign fossil hunters may open up in the future.

Stephen Brusatte and Hans-Dieter Sues are some of the palaeontologists who have been working in Central Asia, where a series of former Soviet nations have almost as much potential as Mongolia to yield huge numbers of new dinosaur fossils. A 90-million-year-old horse-sized tyrannosaur from the Kyzylkum Desert of northern Uzbekistan was named as *Timurlengia euotica* (after a 14th-century Central Asian warlord) in early 2016.[1] It helped fill a 20-million-year gap in the fossil record that was obscuring the origins of *T. rex* itself, and the fossil included a brain case that showed the species had a relatively large brain with an elongated ear canal that helped it hear low-frequency sounds. Our knowledge of the tyrannosaur lineage has expanded enormously in recent years, with discoveries of new species right across Asia. Neighbouring Kazakhstan has recently yielded part of the upper jaw of a huge tyrannosaur, replete with numerous teeth.

Myanmar has also produced some truly breathtaking new specimens. Here, a team led by Xing Lida of the China University of Geosciences in Beijing has been searching for fossils in the Hukawng Valley in the northern state of Kachin. In late 2014, after they began to find dinosaur and bird feathers preserved in 99-million-year-old amber in jewellery markets, they started to prospect directly at amber quarries. They have since recovered many important specimens exquisitely preserved in amber, including two feathered bird or dinosaur wings with claws; a probable dinosaur tail with muscles, skin and feathers; and many nearly complete birds, mammals and snakes.

Some parts of the world you'd expect to have already been

thoroughly picked over for fossils are continuing to turn up new specimens. North America falls obviously into this category, but a series of ceratopsids from there is now described every year. New finds are also coming from Western Europe, particularly Portugal and Spain.

Several exciting new finds have been made in the United Kingdom in recent years. Following spring storms in early 2014, brothers and amateur fossil hunters Nick and Rob Hanigan made a thrilling find on Lavernock beach in the Vale of Glamorgan, southern Wales. Following the collapse of some cliffs, a series of blocks with unusual bones had been exposed. Hours of work yielded five slabs of rock with animal bones, which the pair carried back to their car. They spent almost a year carefully preparing the specimen to uncover as much as possible, before donating it to the National Museum of Wales in Cardiff.

The species was described as *Dracoraptor* by a team including Dave Martill in 2016, and at 200 million years old may be the earliest Jurassic dinosaur known in the world.[2] *Dracoraptor* was an agile carnivore, 2 metres long with a long tail and just half a metre tall. Related to a small North American dinosaur called *Coelophysis*, it likely fed upon insects and small mammals and reptiles. It is only the second dinosaur ever found in Wales, and the nation's first carnivore. Dinosaurs were evolving rapidly in the early Jurassic, forming some of the major groups, but there are few specimens from this time worldwide, so the new fossil could help answer important questions.

From the opposite end of the United Kingdom, on Scotland's remote Isle of Skye come hundreds of dinosaur footprints, trackways left along the shore of a brackish lagoon by sauropods 161 million years ago in the Middle Jurassic. Described in 2015 by a team led by Brusatte of the University of Edinburgh, it is not only the largest dinosaur site in Scotland, but also provides important data from a period of dinosaur evolution that is very poorly understood.[3] The new tracksite from Skye is one of the most remarkable dinosaur discoveries ever made in Scotland', Brusatte told reporters. 'There are so many tracks crossing each other that it looks like a dinosaur disco preserved in stone.'

Massospondylus from South Africa was one of the very first dinosaurs named, by Richard Owen at the Natural History Museum in London in 1854. But South Africa is a location where many new Early Jurassic discoveries are now being made, the work spearheaded by Jonah Choiniere at the Evolutionary Studies Institute of the University of Witwatersrand in Johannesburg. These include 10 nests of possible *Massospondylus* eggs found in a cliff at the Golden Gate Highlands National Park of the eastern Free State, which, at 190 million years old, may be the oldest dinosaur nests ever found.[4] In 2015, a new sauropodomorph dinosaur, a relative of *Massospondylus*, was named *Sefapanosaurus* (meaning 'cross lizard', in reference to the shape of its ankle, in the Sesotho language of Zastron region in which it was found).[5]

Further north in Africa, in Tanzania, a giant titanosaur sauropod called *Rukwatitan bisepultus* was described by palaeontologists including Ohio University's Eric Gorscak in 2015.[6] The finding added to the handful of titanosaurs found so far on the African mainland. 'Much of what we know regarding titanosaurian evolutionary history stems from numerous discoveries in South America – a continent that underwent a steady separation from Africa during the first half of the Cretaceous Period', Gorscak told reporters. 'With the discovery of *Rukwatitan* and study of the material in nearby Malawi, we are beginning to fill a significant gap from a large part of the world.'

This just gives you a taste of the huge numbers of finds being made right now and from nearly every region of the world. There are plenty of others besides and, no doubt, like me you will be watching the field with great interest.

GLOSSARY

ankylosaurs – armoured ornithischian dinosaurs that were built like tanks and covered in bony plates and spikes. Some had tail clubs.

azhdarchid pterosaurs – a family of huge pterosaurs from the Late Cretaceous including the largest flying animals of all time, such as *Hatzegopteryx* and *Quetzalcoatlus*.

badlands – a type of terrain where extensive erosion of softer sedimentary rocks and clay-rich soils by wind and water creates uneven ground with steep slopes, minimal vegetation and dryness. They are so named because they are difficult to traverse.

bipedal – two-legged, rather than four-legged.

carnosaurs – predatory theropod dinosaurs, including *Allosaurus* and carcharodontosaurs such as *Carcharodontosaurus*, *Giganotosaurus* and *Tyrannotitan*.

ceratopsian – wider group of horned ornithischian dinosaurs including *Triceratops* and its kin, and earlier less specialised members of the group such as *Psittacosaurus* and *Protoceratops*.

ceratopsid – the group of large horned ceratopsians related to *Triceratops* and found largely in western North America in the Late Cretaceous.

coelurosaurs – a group of theropod dinosaurs related to birds. It includes compsognathids, tyrannosaurs, ornithomimosaurs and maniraptorans.

comparative anatomy – the study of the physical features of species in comparison with one another. Allows experts to draw inferences about extinct species and 'reconstruct' them based on living ones.

Cretaceous – the geological period beginning 145 million years ago and ending 66 million years ago.

deciduous – a word used to describe plants that lose their leaves in the cold season.

describe – in the context of discovery of new species the act of 'describing' or a 'description' means the formal task of naming and defining the characteristics of a novel plant or animal.

dromaeosaurs – a group of small bird-like theropods (including *Velociraptor*) that were fast runners, effective predators and perhaps even pack hunters.

endocast – an internal cast of a hollow object, such as a brain case.

epiossification – small bony extensions that form bumps or ridges around the edge of another bone, usually a ceratopsian neck frill.

field jacket – a plaster cast around a fossil embedded in the rock it was found in – used to protect it during transport.

formation – layers of rock laid down, without breaks to divide them, over a relatively short period of geological time.

genome – the entire sum of an organism's DNA.

Gondwana – southern supercontinent that began to break up into Africa, the Arabian Peninsula, Australia, South America, India, Madagascar and Antarctica 184 million years ago.

Glossary

hadrosaurs – the herbivorous duck-billed dinosaurs, a group of ornithischians that included *Maiasaura* and *Parasaurolophus*.
ichnofossils – fossils revealing traces of life rather than the life form itself, such as dinosaur footprints.
integumentary – structures originating from the skin of an organism, such as fluff, scales, fur and feathers.
island rule – a theory that when mainland animals colonise islands, small species tend to evolve larger bodies, and large species tend to evolve smaller bodies.
Jurassic – the geological period beginning 201 million years ago and ending 145 million years ago.
keratin – the protein that makes up nails and claws, reptile scales, mammal fur and bird (and dinosaur) feathers.
Laurasia – northern supercontinent that began to break up into North America, Greenland and Eurasia (not including India) about 60 million years ago.
maniraptors – theropod dinosaurs most tightly linked to birds. They include dromaeosaurs, oviraptorosaurs and therizinosaurs.
megaraptorids – a group of Southern Hemisphere carnivorous theropods with large claws, which evolved on Gondwana and have been found in places such as South America and Australia.
Mesozoic – the geological era comprising the Triassic, Jurassic and Cretaceous periods.
morphology – the form, shape and structure of an organism or physical feature.
ornithischian – one of the two divisions within the dinosaur group; the other is the saurischians. The ornithischian or 'bird-hipped' dinosaurs, include heavy-set and armoured species and herd-living herbivores, such as the hadrosaurs.
oviraptorosaurs – a group of parrot-beaked omnivorous theropods with short pygostyle tails, that are inferred to have had feather fans attached for display.
palaeontology – The study of prehistoric life, based on the fossils of animals, plants and other organisms and the age and details of the layers of rock they are found in.
Pangaea – the supercontinent made up of all the modern-day continents, which began to break up into Laurasia and Gondwana 200 million years ago.
pennaceous – (*pen-ay-SHUHS*) a word used to describe feathers that are the typical modern feather shape, with a central shaft running along their length and interlocking barbs running off to either side.
Permian – the geological period beginning 299 million years ago and ending 251 million years ago. It ended with the largest mass extinction in the earth's history and preceded the Triassic.
pneumatisation – invasion of air sacs into bird bones creating a lightweight honeycomb structure.
prosauropod – early relatives of the sauropods, which include some of the earliest of all known dinosaurs. These animals, including *Massospondylus*, *Glacialisaurus* and *Lufengosaurus*, had long necks and walked some of the time on two legs.

protofeather – simple filament representing one of the earliest stages of feather evolution.
pygostyle – the shortened tail structure of modern birds, to which a fan of feathers attaches.
quill knobs – pits where the ligaments of flight feathers attach to the arm bones of modern birds.
saurischian – one of the two divisions within the dinosaur group; the other is the ornithiscians. The saurischian or 'bird-hipped' dinosaurs, include long-necked sauropods and all the bipedal, predatory theropods.
sauropod – a diverse group of massive, long-necked herbivorous dinosaurs that included the largest land animals that ever lived. Together with the theropods, they form the saurischian branch of the dinosaur family.
sauropodomorph – sauropods and their relatives and ancestral forms, including the prosauropods.
sexual dimorphism – when the male and female of a species are clearly differentiable in size and/or appearance.
sexual selection – a process of natural selection whereby certain characteristics in one sex are chosen preferentially by the other and therefore perpetuated in succeeding generations.
soft tissue – any fossilised body part not created by the remains of bones ('hard tissue'), such as muscles, skin, internal organs, fur and feathers.
sternum – a bone between the ribs that acts like a keel and anchors the breast muscles that power the wings in modern birds.
stratigraphy – the study of the order and relative position of rock layers and how they relate to geological time.
taxonomy – biological discipline concerned with the classification and naming of organisms.
terrestrial – living on land.
theropods – a large group of bipedal saurischian dinosaurs that included all of the carnivorous species, such as *T. rex*, *Allosaurus*, *Sinosauropteryx* and all birds.
therizinosaurs – (*ther-uh-ZIN-oh-SOREs*) carnivorous theropods whose teeth suggest they had returned to a vegetarian diet. The name means 'reaping' or 'scything' lizard.
titanosaurs – largely Southern Hemisphere sauropods, which diversified on Gondwana and reached a zenith in size in Patagonia in the Cretaceous Period.
Triassic – the geological period beginning 252 million years ago and ending 201 million years ago.
troodontids – a group of small (less than 100-kilogram) theropods related to dromaeosaurs and birds. Species include *Troodon*, *Anchiornis*, *Xiaotingia* and *Mei*.
warm-blooded – animals with high metabolic rates that can regulate their internal body temperature to within a narrow range, a process known as thermoregulation. Also called endothermic.

FURTHER READING

Gangloff, Roland A, *Dinosaurs Under the Aurora*, Indiana University Press, Bloomington, 2012

Nothdurft, William with Josh Smith, *The Lost Dinosaurs of Egypt*, Random House, New York, 2002

Novacek, Michael, *Dinosaurs of the Flaming Cliffs*, Anchor Books, New York, 1997

Pickrell, John, *Flying Dinosaurs: How Fearsome Reptiles Became Birds*, NewSouth, Sydney, 2014

Ryan, Michael J, Brenda J Chinnery-Allgeier & David A Eberth (eds), *New Perspectives on Horned Dinosaurs: The Royal Tyrell Museum Ceratopsian Symposium*, Indiana University Press, Bloomington, 2010

Smith, Elizabeth, *Black Opal Fossils of Lightning Ridge: Treasures from the Rainbow Billabong*, Kangaroo Press, Sydney, 1999

Stilwell, Jeffrey D & John A Long, *Frozen in Time: Prehistoric Life in Antarctica*, CSIRO Publishing, Melbourne, 2011

Weishempel, David B & Coralia-Maria Jianu, *Transylvanian Dinosaurs*, Johns Hopkins University Press, Baltimore, 2011

ACKNOWLEDGMENTS

Where to start with thanking everybody who had a hand in making this book happen? I'd like to acknowledge all the experts who gave their time to talk to me over the phone or in person, endured my numerous emails, sent me journal articles and suggestions, allowed me to use their artworks, answered subsequent questions, and reviewed and commented on drafts of my chapters. These include: Andry Atuchin, Paul Barrett, Phil Bell, Mike Benton, Davide Bonadonna, Jenni Brammall, Caleb Brown, Luis Chiappe, Julius Csotonyi, Phil Currie, Kristi Curry Rogers, Pat Druckenmiller, Gareth Dyke, Gregory Erickson, Federico Fanti, Andy Farke, Tony Fiorillo, Pascal Godefroit, Peter Hews, Nizar Ibrahim, Tsogtbaatar Khishigjav, Dave Martill, Michael Pittman, Diego Pol, Michael Ryan, Steve Salisbury, Joe Sertich, Elizabeth Smith, Nathan Smith, Corwin Sullivan, David Weishampel, Emily Willoughby, Lida Xing and Xu Xing.

I'd like to thank the State Library of New South Wales in Sydney for providing both books and a place for quiet contemplation and study. I'm grateful to NewSouth Publishing and Columbia University Press for supporting my concept for this book and helping me refine my ideas, and Graham Price for helping with my website for this and my previous book, *Flying Dinosaurs*. I'd also like to thank Carolyn Reynolds for transcribing some of my interviews; Anthony Chapman for spending many hours reading over the draft chapters and providing comments for revisions; and Elspeth Menzies, Emily Stewart and Nicola Young for commenting on, editing and helping to improve the copy. Finally, I'm grateful to my family for being proud of me and encouraging and supporting me in all of my scientific and journalistic endeavours.

NOTES

Introduction: A new golden age for dinosaur science

1. Latin names, such as *Tyrannosaurus rex* or *Homo sapiens*, have two parts. The first refers to the genus, which is shared by a closely related group of species, while the second refers to the species in question. The vast majority of dinosaur genera are 'monotypic', meaning they contain just a single species, so in common parlance the generic name, such as *Tyrannosaurus* or *Triceratops*, is often used to refer to the species. Throughout this book I use that convention without spelling out the full name of each species, unless it's necessary for clarity.
2. P Dodson, 'Counting dinosaurs: how many kinds were there?', *Proceedings of the National Academy of Sciences*, 1990, vol. 87, pp. 7608–12, <www.pnas.org/content/87/19/7608.full.pdf>.
3. H Fountain, 'Many more dinosaurs still to be found', *New York Times*, 12 September 2006, <www.nytimes.com/2006/09/12/science/12observ.html?_r=2&>.
4. A widely trusted list is the one kept by amateur palaeontologist George Olshevsky at <www.polychora.com/dinolist.html>, although the figures vary slightly from those on another regularly updated list on Wikipedia, <en.wikipedia.org/wiki/List_of_dinosaur_genera>.
5. For many decades an extinct, long-necked relative of the rhinoceros – an indricothere, *Paracetherium* – was thought to be to be the biggest ever land mammal. These Oligocene (from 23–34 million years ago) animals weighed up to 20 tonnes and were up to 7.8 metres long. But a 2015 study suggested than an extinct elephant, *Palaeoloxodon*, may have been even bigger, weighing 21 tonnes or more.

1 Monster from the Cretaceous lagoon

1. The fossil collector Richard Markgraf was originally from Bohemia, today part of the Czech Republic, but lived for many years in Egypt, working as a *sammler* or collector for fossil hunters, including Henry Fairfield Osborn of the American Museum of Natural History in New York. Markgraf was talented at finding and excavating fossils, and had several species named in his honour, such as the primate *Moeripthecus markgrafi* and the fish *Markgrafica libica*.
2. Pronounced *kahr-KAR-o-DON-to-SOR-us*.
3. F Therrien & DM Henderson, 'My theropod is bigger than yours… or not: estimating body size from skull length in theropods', *Journal of Vertebrate Paleontology*, 2007, vol. 27, no. 1, pp. 108–15.
4. JB Smith et al., 'A giant sauropod dinosaur from an Upper Cretaceous mangrove deposit in Egypt', *Science*, 2001, vol. 292, no. 5522, pp. 1704–06.

2 All hail the dino-bat

1. Dinosaurs are made up of two major subgroups: the saurischian or 'lizard-hipped' dinosaurs, which included the giant long-necked sauropods (such as *Diplodocus*) and all the bipedal, predatory theropods (such as *Tyrannosaurus* and *Velociraptor*);

and the ornithischian or 'bird-hipped' dinosaurs, which included heavy-set and armoured species (such as *Triceratops* and *Ankylosaurus*) and herd-living herbivores (such as *Hadrosaurus* and *Pachycephalosaurus*). Officially, dinosaurs are deemed to be all the animals that descended from the last shared ancestor of the ornithischian and saurischian groups. Confusingly, birds are theropods and are therefore part of the 'lizard-hipped' group.

2   Pronounced *tee-ANN-you-long*.
3   Pronounced *ANG-kee-OR-niss* and *shyow-TIN-gee-uh*.
4   D Levin, 'The murky origins of the largest dinosaur museum in the world', *New York Times*, 2 December 2013, <www.nytimes.com/2015/12/03/world/asia/the-murky-origins-of-the-largest-dinosaur-museum-in-the-world.html>.
5   Pronounced *ee chee*.
6   K Padian, 'Palaeontology: dinosaur up in the air', *Nature*, 2015, vol. 521, no. 7550, pp. 40–41.
7   E Yong, 'Chinese dinosaur had bat-like wings and feathers', Not Exactly Rocket Science blog, *National Geographic*, 29 April 2015, <phenomena.nationalgeographic.com/2015/04/29/chinese-dinosaur-had-bat-like-wings-and-feathers>.
8   Pronounced *SIGH-oor-uh-MEE-mus*.
9   The story of China's feathered dinosaurs, the history of their discovery, the origins of flight and the problem of smuggled and fake fossils are subjects I talk about at length in my previous book *Flying Dinosaurs: How Fearsome Reptiles Became Birds*, so that's where to find more information, if you're interested in this topic.
10 G Han et al., 'A new raptorial dinosaur with exceptionally long feathering provides insights into dromaeosaurid flight performance', *Nature Communications*, 2014, vol. 5, article no. 4382, <www.nature.com/ncomms/2014/140715/ncomms5382/full/ncomms5382.html>.
11 J Lü & S Brusatte, 'A large, short-armed, winged dromaeosaurid (Dinosauria: Theropoda) from the Early Cretaceous of China and its implications for feather evolution', *Scientific Reports*, 2015, vol. 5, article no. 11775, <www.nature.com/articles/srep11775>.
12 B Bryson, *A Short History of Nearly Everything*, Broadway Books, New York, 2003.
13 G Vince, *Adventures in the Anthropocene: A Journey to the Heart of the Planet We Made*, Chatto & Windus, London, 2014.
14 'Canadian researchers discover fossils of first feathered dinosaurs from North America', Dinosaur Paleontology in Geoscience, University of Calgary, n.d., <www.ucalgary.ca/drg/newsmakers/discoveries/feathered.ornithomimus>.
15 C Sullivan et al., 'The vertebrates of the Jurassic Daohugou Biota of northeastern China', *Journal of Vertebrate Paleontology*, 2014, vol. 34, no. 2, pp, 243–80, <www.tandfonline.com/doi/abs/10.1080/02724634.2013.787316?journalCode=ujvp20>.

3   Dwarf dinosaurs and trailblazing aristocrats

1   M Simon, 'Absurd creature of the week: the 16-foot-tall reptilian stork that delivered death instead of babies', *Wired*, 1 November 2013, <www.wired.com/2013/11/absurd-creature-of-the-week-quetz>.
2   Quoted in DB Weishampel & C-M Jianu, *Transylvanian Dinosaurs*, Johns Hopkins University Press, Baltimore, 2011, p. 10.
3   See J Pickrell, *Flying Dinosaurs: How Fearsome Reptiles Became Birds*, NewSouth, Sydney, 2014, pp. 38–46.

Notes

4   R Elsie, 'The photo collection of Bajazid Doda', Early Photography in Albania, <www.albanianphotography.net/doda>.
5   Weishampel & Jianu, *Transylvanian Dinosaurs*, p. 8.
6   R Elsie, *Historical Dictionary of Albania*, 2nd edn, Scarecrow Press, Lanham, Maryland, 2010, p. 332.
7   R Elsie, '1907. Baron Franz Nopcsa: the baron held hostage in the mountains of Dibra', Texts and Documents of Albanian History, <albanianhistory.net/1907_Nopcsa2/index.html>.
8   Quoted in D Grigorescu, 'The latest Cretaceous fauna with dinosaurs and mammals from the Haţeg Basin: a historical overview', *Palaeogeography, Palaeoclimatology, Palaeoecology*, 2010, vol. 293, nos 3–4, pp. 271–82.
9   DB Weishampel & O Kerscher (trans.), 'Franz Baron Nopcsa by András Tasnádi Kubacska', *Historical Biology*, 2013, vol. 25, no. 4, pp. 391–544.
10 D Naish, *The Great Dinosaur Discoveries*, University of California Press, Berkeley, 2009.
11 A Wilkins, 'The last mammoths died out just 3600 years ago ... but they should have survived', io9, 25 March 2012, <io9.com/5896262/the-last-mammoths-died-out-just-3600-years-agobut-they-should-have-survived>.
12 PM Sander et al., 'Bone histology indicates insular dwarfism in a new Late Jurassic sauropod dinosaur', *Nature*, 2006, vol. 441, no. 7094, pp. 739–41.
13 A Otero, 'The appendicular skeleton of *Neuquensaurus*, a Late Cretaceous saltasaurine sauropod from Patagonia, Argentina', *Acta Palaeontologica Polonica*, 2010, vol. 55, no. 3, pp. 399–426.
14 JS Marpmann et al., 'Cranial anatomy of the Late Jurassic dwarf sauropod *Europasaurus holgeri* (Dinosauria, Camarasauromorpha): ontogenetic changes and size dimorphism', *Journal of Systematic Palaeontology*, 2015, vol. 13, no. 3, pp. 221–63, <www.tandfonline.com/doi/full/10.1080/14772019.2013.875074>.
15 Pronounced *ba-LAH-wuhr*.
16 Z Csiki et al., 'An aberrant island-dwelling theropod dinosaur from the Late Cretaceous of Romania', *Proceedings of the National Academy of Sciences*, 2010, vol. 107, no. 35, pp. 15357–61, <www.pnas.org/content/107/35/15357.full>.
17 A Cau et al., 'The phylogenetic affinities of the bizarre Late Cretaceous Romanian theropod *Balaur bondoc* (Dinosauria, Maniraptora): dromaeosaurid or flightless bird?', *PeerJ*, 2015, vol. 3, article no. e1032.

4   Horny ornaments and sexy ceratopsians

1   Copy of letter, A Smith Woodward to C Sternberg, 11 January 1918, Natural History Museum Archives, London.
2   AA Farke, et al., 'A new centrosaurine from the Late Cretaceous of Alberta, Canada, and the evolution of parietal ornamentation in horned dinosaurs', *Acta Palaeontologica Polonica*, 2011, vol. 56, no. 4, pp. 691–702, <www.app.pan.pl/archive/published/app56/app20100121.pdf>.
3   P Dodson, 'Forty years of Ceratophilia', in MJ Ryan et al. (eds), *New Perspectives on Horned Dinosaurs: The Royal Tyrrell Museum Ceratopsian Symposium*, Indiana University Press, Bloomington, 2010, p. 12.
4   SD Sampson et al., 'New horned dinosaurs from Utah Provide evidence for intracontinental dinosaur endemism', *PLOS ONE*, 2010, vol. 5, no. 9, article no. e12292, <journals.plos.org/plosone/article?id=10.1371/journal.pone.0012292>.

5   SD Sampson et al., 'A remarkable short-snouted horned dinosaur from the Late Cretaceous (late Campanian) of southern Laramidia', *Proceedings of the Royal Society B: Biological Sciences*, 2013, vol. 280, no. 1766, article no. 20131186, <rspb.royalsocietypublishing.org/content/280/1766/20131186>.

6   DC Evans, 'Cranial anatomy of *Wendiceratops pinhornensis* gen. et sp. nov., a Centrosaurine Ceratopsid (Dinosauria: Ornithischia) from the Oldman Formation (Campanian), Alberta, Canada, and the evolution of Ceratopsid nasal ornamentation', *PLOS ONE*, 2015, vol. 10, no. 7, article no. e0130007, <journals.plos.org/plosone/article?id=10.1371/journal.pone.0130007>.

7   AA Farke, 'Evidence of combat in *Triceratops*', *PLOS ONE*, 2009, vol. 4, no. 1, article no. e4252, <http://journals.plos.org/plosone/article?id=10.1371/journal.pone.0004252>.

8   Pronounced *SIT-a-ko-SOR-us*.

9   D Naish, 'Dinosaurs and their exaggerated structures: species recognition aids, or sexual display devices?', Tetrapod Zoology blog, *Scientific American*, <blogs.scientificamerican.com/tetrapod-zoology/species-recognition-vs-sexual-selection-in-dinosaurs>.

10 C Brown, '"Hellboy" the *Regaliceratops*: what this new horned dinosaur tells us about the evolution of Ceratopsians', Royal Tyrrell Museum Speaker Series, at <www.youtube.com/watch?v=_VcxShseKzo>.

11 CM Brown & DM Henderson, 'A new horned dinosaur reveals convergent evolution in cranial ornamentation in Ceratopsidae', *Current Biology*, 2015, vol. 25, no. 12, pp. 1641–48, <www.cell.com/current-biology/abstract/S0960-9822%2815%2900492-3>.

12 S Hudes, 'What's it like to discover a dinosaur?', *Calgary Herald*, 9 October 2015, <calgaryherald.com/news/local-news/whats-it-like-to-discover-a-dinosaur-species>.

5   The 'unusual terrible hands'

1   TR Holtz Jr, 'Palaeontology: mystery of the horrible hands solved', *Nature*, 2014, vol. 515, no. 7526, pp. 203–205.

2   Z Kielan-Jaworowska & N Dovchin, 'Narrative of the Polish-Mongolian palaeontological expeditions 1963–1965', *Palaeontologia Polonica*, 1969, vol. 19, pp. 9–30, <palaeontologia.pan.pl/Archive/1968-19_7-30_1-4.pdf>.

3   H Osmólska & E Roniewicz, 'Deinocheiridae, a new family of theropod dinosaurs', *Palaeontologia Polonica*, 1970, vol. 21, pp. 5–19, <palaeontologia.pan.pl/Archive/1969-21_5-22_1-5.pdf>.

4   RL Cifelli et al., 'In memoriam: Zofia Kielan-Jaworowska (1925–2015)', *Acta Palaeontologica Polonica*, 2015, vol. 60, no. 2, pp 287–90, <www.app.pan.pl/archive/published/app60/app001832015.pdf>.

5   Z Kielan-Jaworowska, 'Autobiografia', *Nadbitka z Kwartalnika Historii Nauki i Techniki*, 2005, vol. 50, no. 1, pp. 7–50, trans. Kielan-Jaworowska, <paleoglot.org/files/ZKJ%20autobio2.doc>.

6   PJ Currie et al., 'Hands, feet, and behaviour in *Pinacosaurus* (Dinosauria: Ankylosauridae)', *Acta Palaeontologica Polonica*, 2011, vol. 56, no. 3, pp. 489–504, <www.app.pan.pl/archive/published/app56/app20100055.pdf>.

7   PR Bell et al., 'Tyrannosaur feeding traces on *Deinocheirus* (Theropoda: ?Ornithomimosauria) remains from the Nemegt Formation (Late Cretaceous),

Notes

Mongolia', *Cretaceous Research*, 2012, vol. 37, pp. 186–90.
8   Y-N Lee et al., 'Resolving the long-standing enigmas of a giant ornithomimosaur *Deinocheirus mirificus*', *Nature*, 2014, vol. 515, no. 7526, pp. 257–60.
9   K Condon, 'Currie part of international team to solve mystery of bizarre Mongolian dinosaur', University of Alberta, 22 October 2014, <uofa.ualberta.ca/news-and-events/newsarticles/2014/october/currie-part-of-international-team-to-solve-the-mystery-of-bizarre-mongolian-dinosaur>.

6   Scandalous behaviour and enfluffled vegetarians

1   'An amazing collection of dinosaur remains found in volcanic ash in Siberia', *Siberian Times*, 29 August 2012, <siberiantimes.com/science/casestudy/news/an-amazing-collection-of-dinosaur-remains-found-in-volcanic-ash-in-siberia>.
2   Pronounced *KOO-lin-dah-DROH-mee-us*.
3   'Feather-like structures and scales in a Jurassic neornithischian dinosaur from Siberia', supplement to the online *Journal of Vertebrate Paleontology*, October 2013, p. 135, <http://vertpaleo.org/Annual-Meeting/Future-Past-Meetings/MeetingPdfs/SVP-2013-merged-book-10-15-2013.aspx>.
4   VR Alifanov, 'The discovery of Late Jurassic dinosaurs in Russia', *Doklady Earth Sciences*, 2014, vol. 455, no. 2, pp. 365–67.
5   VR Alifanov & SV Saveliev, 'Two new ornithischian dinosaurs (Hypsilophodontia, Ornithopoda) from the Late Jurassic of Russia', *Paleontological Journal*, 2014, vol. 48, no. 4, pp. 414–25.
6   'Re: Kulindapteryx and Daurosaurus, new hypsilophodont ornithopods from Upper Jurassic of Siberia, Russia', Dinosaur Mailing List, 5 July 2014, <dml.cmnh.org/2014Jul/msg00028.html>.
7   P Godefroit, SM Sinitsa et al., 'A Jurassic ornithischian dinosaur from Siberia with both feathers and scales', *Science*, 2014, vol. 345, no. 6195, pp. 451–55.
8   VR Alifanov & SV Saveliev, 'The most ancient ornithomimosaur (Theropoda, Dinosauria), with cover imprints from the Upper Jurassic of Russia', *Paleontological Journal*, 2015, vol. 49, no. 6, pp. 636–50.
9   B Conmy, 'Research by UCC professors suggests dinosaurs were feathered', UCC Express, 19 August 2014, <uccexpress.ie/new-research-by-ucc-professor-suggests-all-dinosaurs-were-feathered-brian-conmy>.
10 M Balter, 'Earliest dinosaurs may have sported feathers', *Science* online news, 24 July 2014, <www.sciencemag.org/news/2014/07/earliest-dinosaurs-may-have-sported-feathers>.
11 P Ghosh, '"Fluffy and feathery" dinosaurs were widespread', BBC News, 25 July 2014, <www.bbc.com/news/science-environment-28407381>.
12 'Fossils found in Siberia suggest all dinosaurs had feathers', University of Bristol News, 24 July 2014, <www.bristol.ac.uk/news/2014/july/fossils-found-in-siberia-suggest-all-dinosaurs-had-feathers.html>.
13 A Liesowska, 'Hello to the Sibirosaurus? New dinosaur discovered by university scientists', *Siberian Times*, 2 March 2015, <siberiantimes.com/science/casestudy/news/n0167-hello-to-the-sibirosaurus-new-dinosaur-discovered-by-university-scientists>.
14 AO Averianov et al., 'A sauropod foot from the Early Cretaceous of western Siberia, Russia', *Acta Palaeontologica Polonica*, 2002, vol. 47, no. 1, pp. 117–24.

15 AV Lopatin, et al., 'A unique burial site of Early Cretaceous vertebrates in Western Siberia (the Shestakovo 3 locality, Kemerovo Province, Russia)', *Doklady Biological Sciences*, 2015, vol. 462, no. 1, pp. 148–51.
16 P Nechayeva, 'Siberian scientists checking if they found a 130 million year old horned dinosaur "nuclear family"', *Siberian Times*, 7 July 2014, <siberiantimes.com/science/others/news/siberian-scientists-checking-if-they-have-found-a-130-million-year-old-horned-dinosaur-nuclear-family>.
17 RA Gangloff, *Dinosaurs Under the Aurora*, Indiana University Press, Bloomington, 2012, p. 23.
18 P Godefroit et al., 'The last polar dinosaurs: high diversity of latest Cretaceous arctic dinosaurs in Russia', *Naturwissenschaften*, 2009, vol. 96, no. 4, pp. 495–501.
19 R Gray, 'Dinosaurs could survive cold conditions', *Daily Telegraph* (London), 24 January 2009, <www.telegraph.co.uk/news/earth/earthnews/4330218/Dinosaurs-could-survive-cold-conditions.html>.

7 Cretaceous creatures of the frozen north
1 R Fiorillo et al., 'Herd structure in Late Cretaceous polar dinosaurs: a remarkable new dinosaur tracksite, Denali National Park, Alaska, USA', *Geology*, 2014, vol. 42, no. 8, pp. 719–22, <geology.gsapubs.org/content/42/8/719>.
2 AR Fiorillo & TL Adams, 'A therizinosaur track from the lower Cantwell Formation (Upper Cretaceous) of Denali National Park, Alaska', *Palaios*, 2012, vol. 27, no. 6, pp. 395–400.
3 AR Fiorillo et al., 'Dinosauria and fossil Aves footprints from Denali's lower Cantwell Formation', *Alaska Park Science*, 2007, vol. 6, no. 2, 'Crossing Boundaries in a Changing Environment', pp. 41–42, <https://www.nps.gov/articles/aps-v6-i2-c9.htm>.
4 AR Fiorillo & RS Tykoski, 'A new Maastrichtian species of the centrosaurine ceratopsid *Pachyrhinosaurus* from the North Slope of Alaska', *Acta Palaeontologica Polonica*, 2012, vol. 57, no. 3, pp. 561–73, <www.app.pan.pl/archive/published/app57/app20110033.pdf>.
5 H Mori et al., 'A new Arctic hadrosaurid from the Prince Creek Formation (lower Maastrichtian) of northern Alaska', *Acta Palaeontologica Polonica*, 2016, vol. 61, no. 1, pp. 15–32, <www.app.pan.pl/article/item/app001522015.html>.
6 K Haughney, 'It's a lost world: researchers discover new dinosaur in Arctic', Florida State University, 22 September 2015, <www.fsu.edu/indexTOFStory.html?lead.lostworld>.
7 AR Fiorillo & RS Tykoski, 'An immature *Pachyrhinosaurus perotorum* (Dinosauria: Ceratopsidae) nasal reveals unexpected complexity of craniofacial ontogeny and integument in *Pachyrhinosaurus*', *PLOS ONE*, 2013, vol. 8, no. 6, article no. e65802, <journals.plos.org/plosone/article?id=10.1371/journal.pone.0065802>.
8 AR Fiorillo & RS Tykoski, 'A diminutive new tyrannosaur from the top of the world', *PLOS ONE*, 2014, vol. 9, no. 3, article no. e91287, <journals.plos.org/plosone/article?id=10.1371/journal.pone.0091287>.
9 X Xu et al., 'First ceratopsid dinosaur from China and its biogeographical implications', *Chinese Science Bulletin*, 2010, vol. 55, no. 16, pp 1631–35.

Notes

8 The hidden treasures Down Under

1. PR Bell et al., 'A large-clawed theropod (Dinosauria: Tetanurae) from the Lower Cretaceous of Australia and the Gondwanan origin of megaraptorid theropods', *Gondwana Research*, Aug 2016, vol. 36, pp. 473–487 <http://www.sciencedirect.com/science/article/pii/S1342937X15002026>.
2. LG Leahey et al., 'Cranial osteology of the ankylosaurian dinosaur formerly known as *Minmi* sp. (Ornithischia: Thyreophora) from the Lower Cretaceous Allaru Mudstone of Richmond, Queensland, Australia', *PeerJ*, 2015, vol. 3, article no. e1475, <peerj.com/articles/1475>.
3. E Smith, *Black Opal Fossils of Lightning Ridge: Treasures from the Rainbow Billabong*, Kangaroo Press, Sydney, 1999.
4. M Archer et al., 'First Mesozoic mammal from Australia – an early Cretaceous monotreme', *Nature*, 1985, vol. 318, no. 6044, pp. 363–66.
5. W Verity, 'Lightning Ridge miner says having a dinosaur named after him will "make him feel old"', ABC News, 16 May 2015, <www.abc.net.au/news/2015-05-16/opal-ridge-miner-discovers-new-dinosaur/6474358>.
6. 'Dinosaur gets a CT scan' (video), UNE School of Environmental and Rural Science, at <www.youtube.com/watch?v=sC4sCjGdlb0>.

9 Record-breaking titans

1. KJ Lacovara et al., 'A gigantic, exceptionally complete titanosaurian sauropod dinosaur from Southern Patagonia, Argentina', *Scientific Reports*, 2014, vol. 4, article no. 6196, <www.nature.com/articles/srep06196>.
2. MP Taylor & MJ Wedel, 'Why sauropods had long necks; and why giraffes have short necks', *PeerJ*, 2013, vol. 1, article no. e36, <peerj.com/articles/36>
3. PM Sander et al., 'Biology of the sauropod dinosaurs: the evolution of gigantism', *Biological Reviews*, 2011, vol. 86, no. 1, pp. 117–55, <onlinelibrary.wiley.com/doi/10.1111/j.1469-185X.2010.00137.x/full>.
4. GD Ruxton & DM Wilkinson, 'The energetics of low browsing in sauropods', *Biology Letters*, 2011, vol. 7, no. 5, pp. 779–81, <rsbl.royalsocietypublishing.org/content/7/5/779>.
5. I Sample, 'Dinosaurs with long necks were like 1950s vacuum cleaners, say scientists', *The Guardian*, 23 March 2011, <www.theguardian.com/science/2011/mar/23/dinosaurs-long-necks-vacuum-cleaners>.
6. HC Fricke, 'Lowland–upland migration of sauropod dinosaurs during the Late Jurassic epoch', *Nature*, 2011, vol. 480, no. 7378, pp. 513–15, <www.nature.com/nature/journal/v480/n7378/full/nature10570.html>.
7. G Grellet-Tinner & LE Fiorelli, 'A new Argentinean nesting site showing neosauropod dinosaur reproduction in a Cretaceous hydrothermal environment', *Nature Communications*, 2010, vol. 1, article no. 32, <www.nature.com/ncomms/journal/v1/n3/full/ncomms1031.html>.
8. LM Chiappe et al., 'Embryonic skulls of titanosaur sauropod dinosaurs', *Science*, 2001, vol. 293, no. 5539, pp. 2444–46.
9. FE Novas et al., 'An enigmatic plant-eating theropod from the Late Jurassic period of Chile', *Nature*, 2015, vol. 522, no. 7556, pp. 331–34.
10. JL Carballido et al., 'A new Early Cretaceous brachiosaurid (Dinosauria, Neosauropoda) from northwestern Gondwana (Villa de Leyva, Colombia)', *Journal of Vertebrate Paleontology*, 2015, vol. 35, no. 5, article no. e980505.

11 MC Langer et al., 'New dinosaur (Theropoda, *stem*-Averostra) from the earliest Jurassic of the La Quinta formation, Venezuelan Andes', *Royal Society Open Science*, 2014, vol. 1, no. 2, article no. 140184, <rsos.royalsocietypublishing.org/content/1/2/140184>.

10 Southern killers set adrift

1 Pronounced *muh-SHE-ka-saurus*.
2 Pronounced *rah-HOON-ah-vis*.
3 S Sampson et al. 'A bizarre predatory dinosaur from the Late Cretaceous of Madagascar', *Nature*, 2001, vol. 409, no. 6819, pp. 504–506.
4 J Roach, 'Dinosaur cannibal: fossil evidence found in Africa', National Geographic News, 28 October 2010, <http://news.nationalgeographic.com/news/2003/04/0402_030402_dinocannibal.html>.
5 Pronounced *foos-ah*.
6 'PaleoPeople: David Krause', The Paleontology Portal, <paleoportal.org/index.php?globalnav=paleopeople&interview_id=3>.
7 'Episode 1: Finding fossils in Madagascar' (podcast), Past Time, <www.pasttime.org/podcast/episode-1-finding-fossils-in-madagascar>.
8 P Dodson, 'Paleontology done right – *Mejungasaurus crenatissimus*', *American Paleontologist*, 2008, vol. 16, no. 2, pp. 26–29, <www.museumoftheearth.org/files/pubtext/item_pdf_198.pdf>.
9 B Handwerk, 'Giant "frog from hell" fossil found in Madagascar', National Geographic News, 18 February 2008, <news.nationalgeographic.com/news/2008/02/080218-giant-frog.html>.
10 B Switek, '10 fossils named for rock stars', Mental Floss, 17 September 2014, <mentalfloss.com/article/58890/10-fossils-named-rock-stars>.
11 MT Carrano et al., 'New materials of *Masiakasaurus knopfleri* Sampson, Carrano, and Forster, 2001, and implications for the morphology of the Noasauridae (Theropoda: Ceratosauria)', *Smithsonian Contributions to Paleobiology*, 2011, no. 95, pp. 1–54, <www.sil.si.edu/smithsoniancontributions/Paleobiology/pdf_hi/SCtP-0095.pdf>.
12 Pronounced *rah-PAY-to-SOR-us*.
13 KC Rogers & CA Forster, 'The last of the dinosaur titans: a new sauropod from Madagascar', *Nature*, 2001, vol. 412, no. 6846, pp. 530–34.
14 S Smith, interview with Kristi Curry Rogers, *Bright Ideas* (radio program), 27 March 2012, <www.mprnews.org/story/2012/03/07/bright-ideas-kristi-curry-rogers>.
15 KC Rogers et al., 'Precocity in a tiny titanosaur from the Cretaceous of Madagascar', *Science*, 2016, vol. 352, no. 6284, pp. 450–53.
16 'Newly discovered baby Titanosaur sheds light on dinosaurs' early lives', National Science Foundation press release 16-046, 21 April 2016, <www.nsf.gov/mobile/news/news_summ.jsp?cntn_id=138275&org=NSF>.
17 'The cold hard numbers', The Bearded Lady Project: Challenging the Face of Science, n.d., <thebeardedladyproject.com/about/the-cold-hard-numbers>.
18 AA Farke & JJW Sertich, 'An abelisauroid theropod dinosaur from the Turonian of Madagascar', *PLOS ONE*, 2013, vol. 8, no, 4, article no. e62047, <journals.plos.org/plosone/article?id=10.1371/journal.pone.0062047>.
19 A Farke, 'Madagascar's lonely little thief', The Integrative Paleontologists, PLOS

blogs, 18 April 2013, <blogs.plos.org/paleo/2013/04/18/dahalokely>.
20 N Drake, 'Fossil from dinosaur era reveals big mammal with super senses', National Geographic News, 5 November 2014, <news.nationalgeographic.com/news/2014/11/141105-mammal-evolution-vintana-fossil-science>.
21 G Filiano, 'Newly discovered fossil is a clue to early mammalian evolution', Stony Brook Newsroom, 5 November 2014, <sb.cc.stonybrook.edu/news/general/141005creature.php>.

11 Polar pioneers and the frozen crested lizard

1 'Archive for Wednesday, January 17th 1912', Scott Polar Research Institute, Department of Geography, University of Cambridge, <www.spri.cam.ac.uk/museum/diaries/scottslastexpedition/1912/01/17>.
2 'Archive for Thursday, February 8th 1912', Scott Polar Research Institute, Department of Geography, University of Cambridge, <www.spri.cam.ac.uk/museum/diaries/scottslastexpedition/1912/02/08/thursday-february-8th-1912>.
3 JN Wilford, 'Bones of dinosaur near South Pole', *New York Times*, 13 March 1991, <www.nytimes.com/1991/03/13/us/bones-of-dinosaur-near-south-pole.html>.
4 WR Hammer & WJ Hickerson, 'A crested theropod dinosaur from Antarctica', *Science*, 1994, vol. 264, no. 5160, pp. 828–30.
5 'Webcast! Exploring Antarctic dinosaurs with the Field Museum', BHL: Notes and News from the BHL Staff, 13 October 2015, <blog.biodiversitylibrary.org/2015/10/live-webcast-today-exploring-antarctic.html>.
6 ND Smith & D Pol, 'Anatomy of a basal sauropodomorph dinosaur from the Early Jurassic Hanson Formation of Antarctica', *Acta Palaeontologica Polonica*, 2007, vol. 52, no. 4, pp. 657–74, <www.app.pan.pl/article/item/app52-657.html>.
7 ND Smith, 'New dinosaurs from the Early Jurassic Hanson Formation of Antarctica, and patterns of diversity and biogeography in Early Jurassic sauropodomorphs', 125th Anniversary Annual Meeting & Expo, Geological Society of America, <gsa.confex.com/gsa/2013AM/webprogram/Paper232923.html>.
8 J Pickrell, 'Two new dinosaurs discovered in Antarctica', National Geographic News, 9 March 2004, <news.nationalgeographic.com/news/2004/03/0309_040309_polardinos.html>.
9 JA Case et al., 'A dromaeosaur from the Maastrichtian of James Ross Island and the Late Cretaceous Antarctic dinosaur fauna', Short Research Paper 083, US Geological Survey and the National Academies, 2007, <pubs.usgs.gov/of/2007/1047/srp/srp083/of2007-1047srp083.pdf>.
10 RC Ely & J Case, 'A re-evaluation of the Early Maastrichtian theropod from James Ross Island, Antarctica, based on pedal morphology', abstract, National Conference on Undergraduate Research, 2015, <ncurdb.cur.org/ncur2015/search/display_ncur.aspx?id=91677>.
11 S Pappas, 'First long-necked dinosaur fossil found in Antarctica', Live Science, 4 November 2011, <www.livescience.com/16883-sauropod-dinosaur-fossil-antarctica.html>.
12 S Rozadilla et al., 'A new ornithopod (Dinosauria, Ornithischia) from the Upper Cretaceous of Antarctica and its palaeobiogeographical implications', *Cretaceous Research*, 2016, vol. 57, pp. 311–24.
13 RA Coria et al., 'A new ornithopod (Dinosauria; Ornithischia) from Antarctica', *Cretaceous Research*, 2013, vol. 41, pp. 186–93.

14 Pronounced *lee-EL-in-a-SOR-a*.
15 HN Woodward et al., 'Growth dynamics of Australia's polar dinosaurs', *PLOS ONE*, 2011, vol. 6, no. 8, article no. e23339, <journals.plos.org/plosone/article?id=10.1371/journal.pone.0023339>.

Future potential
1 SL Brusatte et al., 'New tyrannosaur from the mid-Cretaceous of Uzbekistan clarifies evolution of giant body sizes and advanced senses in tyrant dinosaurs', *Proceedings of the National Academy of Sciences*, 2016, vol. 113, no. 3, pp. 3447–52.
2 DM Martill et al., 'The oldest Jurassic dinosaur: a basal neotheropod from the Hettangian of Great Britain', *PLOS ONE*, 2016, vol. 11, no. 1, article no. e0145713, <journals.plos.org/plosone/article?id=10.1371/journal.pone.0145713>.
3 SL Brusatte & NDL Clark, 'Sauropod dinosaur trackways in a Middle Jurassic lagoon on the Isle of Skye, Scotland', *Scottish Journal of Geology*, 2015, vol. 51, no. 2, pp. 157–64.
4 RR Reisz, 'Oldest known dinosaurian nesting site and reproductive biology of the Early Jurassic sauropodomorph *Massospondylus*', *Proceedings of the National Academy of Sciences*, 2012, vol. 109, no. 7, pp. 2428–33.
5 A Otero et al., 'A new basal sauropodiform from South Africa and the phylogenetic relationships of basal sauropodomorphs', *Zoological Journal of the Linnean Society*, 2015, vol. 174, no. 3, pp. 589–634.
6 E Gorscak et al., 'The basal titanosaurian *Rukwatitan bisepultus* (Dinosauria, Sauropoda) from the middle Cretaceous Galula Formation, Rukwa Rift Basin, southwestern Tanzania', *Journal of Vertebrate Paleontology*, 2014, vol. 34, no. 5, pp. 1133–54.

Credits

All maps are licensed as creative commons from freevectormaps.com. Dinosaur silhouettes with chapter heads are creative commons images sourced from phylopic.org. Credits as follows: Introduction: Raven Amos, Emily Willoughby, Brad McFeeters (vectorised by T Michael Keesey); Egypt: Scott Hartman, Martin Kevil, Dmitry Bogdanov (vectorised by T Michael Keesey); China: Emily Willoughby; Romania: Wikimedia, Scott Hartman, Pete Buchholz; Canada: Craig Dylke, Andrew A Farke; Mongolia: T Michael Keesey; Martin Kevil, FunkMonk; Siberia: Pete Buchholz, Matt Martyniuk (modified by Serenchia), Wikimedia; Alaska: Jaime Headden, Andrew A Farke; Australia: T Tischler, Andrew A Farke, Pete Buchholz; Argentina: Scott Hartman, Jaime Headden, Kenneth Lacovara; Madagascar: Jaime Headden, Scott Hartman, T. Michael Keesey; Antarctica: FunkMonk, Brad McFeeters (vectorised by T Michael Keesey), Scott Hartman; Future potential: Raven Amos, Emily Willoughby, Brad McFeeters (vectorised by T Michael Keesey).

INDEX

ABC radio 148
abelisaurs 172–3, 177, 186
Acta Palaeontologica Polonica 69
Adventures in the Anthropocene (Vince) 42
Aegyptosaurus 15, 17
Aerosteon 136
Africa 4
agitator 144
air pockets in bone 22, 160
Alaska 2–3
    Alaska Peninsula 133
    Alaska Range 115
    Aleutian Islands 133
    Anchorage 122
    Beaufort Sea 121
    Brooks Range 129
    Colville River 121–7
    Denali National Park 115–19
    Fairbanks 117, 127
    Igloo Creek 119
    North Slope 115, 119–25
    Tanana River 120–1
    Yukon River 120–1, 132
Albania xi, 47, 49, 51–4
Alectrosaurus 84
Alexandria 14
Algeria 11, 20, 213
Alifanov, Vladimir 103–105
alligators 22
Allosaurus xi, 16, 136, 172
Alte Akademie, Munich 24
altitude sickness 200
amber 213
American Museum of Natural History, New York 1, 68–9, 84, 87
American Paleontologist 177
ammonites 188, 205
Amundsen, Roald 190
anatomy
    brain 71, 160, 189, 197–8, 213
    claws 80, 136, 146
    ear 213

femur bone 154–7, 187
gastralia 92
lungs 161
neck length 160–1
neural spines 23
nostrils 22, 181
osteoderms 178, 182–3
penis musculature 57
quill knobs 35
sail spines 12, 23
styliform elements 32, 40
webbed feet 22
wings 7, 33, 36
Anchiceratops 78
Anchiornis 27–8, 35, 43, 45, 98, 106
Ancient Egypt 58
angiosperms 205
ankylosaurs 51, 89–90, 112, 132, 138, 194, 207–208
Antarctic Peninsula Paleontology Project 209
Antarctica 3–4, 37, 120, 133, 135, 137–8, 151, 159, 164, 189, 190–211
    Antarctic Peninsula 194, 200, 204, 210
    Beardmore Glacier 191, 194–7
    Gordon Valley 195
    James Ross Island 194, 199, 204, 207
    Mt Buckley 191
    Mt Kirkpatrick 37, 195–203, 210
    Naze Peninsula 205
    Queen Alexandra Range 195
    Seymour Island 194, 209
    Shackleton Glacier 203
    Transantarctic Mountains 195, 200
    Vega Island 194, 205, 208–209
    Weddell Sea 205
Antarctopelta 194, 207
Antetonitrus 159–60
antlers 75

Appalachia 124
Archaeopteryx 43, 98–99
Archaeoraptor 33
Archer, Mike 145
Arctic 4, 112, 115–34
Arctic Ocean 75
Argentina 17
    Auca Mahuevo 165–7, 174
    Auca Mahuida 165
    La Flecha 155–8, 163–4
    La Rioja Province 165
    Neuquén 165
    Patagonia 5, 17, 136–7, 140, 153–69, 184, 193–4, 208, 210
    Sanagasta 165
    Trelew 137, 155–157
Argentine Museum of Natural Science, Buenos Aires 167
Argentinosaurus 49, 154, 156–7, 167
armour plates 18, 61
Augustana College, Illinois 195
Australia 2–4, 133
    Andamooka 142
    Coober Pedy 142
    Dinosaur Cove 140
    Kimberley 140
    Lightning Ridge 134–52
    New South Wales 136–52
    Queensland 137–41, 145, 209
    South Australia 138, 142
    Sydney 1
    Tasmania 56, 193
    Western Australia 140
    White Cliffs 142
Australian Age of Dinosaurs Museum 139
Australian Army 148
Australian Geographic 82, 146
Australian Museum, Sydney 1, 141–2, 144, 148
Australian Opal Centre, Lightning Ridge 139, 145–51
Australovenator 137, 139
Austro–Hungarian Empire 51, 54–5
aye-aye 30
Azhdarchid pterosaurs 49, 61

badlands 67, 92, 179, 186

Bagaceratops 88
Bahariasaurus 15–17, 20
Balaur 61–2
Balkan Wars 54
Barrett, Paul 68–9, 107
Barsbold Rinchen 91
Baryonyx 21
bats 32–3
Bavarian State Collection of Palaeontology and Historical Geology 24
BBC 107
bears 126, 128
Bedouin musicians 18
Beelzebufo 172, 178
Beipiaosaurus 36, 119
Belgium 95
Bell, Phil 89, 91, 95, 137, 147, 149–51
Benton, Mike 49, 58, 104–105, 107
Bering Land Bridge 130, 132
Beurlen, Karl 24
Biology Letters 162
biomechanics 7
birds 8, 32, 41, 43, 49, 74, 79, 102, 115, 120, 130, 136, 144, 161, 175, 178, 180–1, 192, 196–8, 203, 208, 214
    cassowary 75, 198
    chicken 102, 108
    elephant bird 175
    flightless 61
    mallee fowl 155
    megapode 155, 166
    peregrine falcon 125
    scrub turkey 155
Black Opal Fossils of Lightning Ridge (Smith) 141
blizzard 199
blue whale 5
Bolivia 167–8
Bolotsky, Yuri 104
bombing raids 17, 24
Bonaparte, José 167
Borkavic, Ben 78
Bowers, Birdie 192
Brachiosaurus 163, 181
Braddock, Peter 201
Brammall, Jenni 145–51, 184

Index

Brazil 21, 167–68
British Empire 14
Brogan, Rob and Debbie 150
Brown, Caleb 66–7, 76–8
Bruce Museum, Connecticut 34
Brusatte, Stephen 41, 96, 107, 213–14
Bryson, Bill 42
Budapest 53
Buffetaut, Eric 49
Bulgaria 88
bull trout 66
Burma 35

Cage, Nicholas 93
Calgary Herald 78
Camarasaurus 163
Camptosaurus 59
Canada 5, 35
    Alberta 40, 44, 65–78, 91–2, 124, 175
    Calgary 78
    Callum Creek 79
    Castle River 79
    Highwood River 79
    St Marys River 79
    South Saskatchewan River 175
Canadian Museum of Nature, Ottawa 69
Cantwell Formation, Alaska 115
Carabajal, Ariana Paulina 207
Carballido, José Luis 167
Carcharodontosaurus 15–16, 20–1, 25, 164
Carmen Funes Municipal Museum, Plaza Huincul 167, 207
Carnegie, Andrew 67
Carnegie Institution 67
Carpathian Mountains 51, 54–5
Case, Judd 204
centipedes 18
Central Museum of Mongolian Dinosaurs 96
centre of gravity 22
Centrosaurus 73
ceratopsian 2, 5, 36–7, 39, 65–79
    centrosaurines 71, 77
    chasmosaurines 71, 77
    diversity 70, 76, 97

frill epiossifications 70, 72, 77
head ornamentation 5
neck frill 70
ornamentation 70–8
rostral bone 70
Chalk Seas 56
chameleons 72, 74–5, 170, 174, 186
Changyuraptor 6, 36, 41
Chasmosaurus 71
Chiappe, Luis 162–168, 174
Chicxulub Crater, Mexico 113
Chile 167–168
Chilesaurus 168
China 2, 27–46, 71, 76, 96, 100, 108
    Beijing 27, 29, 39
    Gansu Province 39
    Hebei Province 27, 29, 35, 41, 45
    Henan Province 39
    Inner Mongolia Province 29, 35, 39, 41, 45, 83
    Kunming 37
    Liaoning Province 7, 29–30, 35, 39, 45, 99, 106, 109
    Mutoudeng 29
    Pingyi 27, 29, 38
    Qinglong County 29
    Shanxi Province 39
    Sichuan Province 39
    Xinjiang Province 38
    Yunnan Province 37, 39
    Zhejiang Province 39
China University of Geosciences 33, 213
Choiniere, Jonah 215
chrono-species 71, 75
Civic Museum of Natural History, Milan 12
claim-jumping 98, 102
Cleveland Museum of Natural History 69, 212
climate 19
coelacanths 9, 19, 23
Coelophysis 214
cold-blooded 4, 115, 130
colonisation 60
Colorado College 163
communism 114
Congo River, Africa 20

conifer trees 99, 113, 118, 135
Cook, Lieutenant James 193
Cope, Edward Drinker 50, 67, 204
coral reefs 24
Coria, Rodolfo 167, 207, 208
Cosmos 145
Cretaceous, Early 6, 45, 59, 140, 151
Cretaceous, Late 18–20, 38–39, 47, 79, 80, 84, 113, 115–116, 122, 124, 129, 153, 160, 165, 167, 170, 172, 182, 194, 205
Cretaceous–Paleogene extinction 8, 75, 112–3, 209
Crete, Greece 58
crocodiles 9, 11, 15, 19, 22, 57, 61, 79, 110, 130, 138–9, 145, 151, 155, 172, 178, 180, 182, 197
Cryolophosaurus xi, 196–202
Csiki, Zoltán 49, 58, 61
Csotonyi, Julius 74
CT scanning 150, 160, 189, 197
Current Biology 77
Currie, Phil 6, 69, 87, 90–2, 94, 96–7, 147, 198–203, 207
Curry Rogers, Kristina 173, 178–85, 189
cycad palms 135, 205

Dahalokely 172, 185, 186–7
Daohugou Biota 45
Darfefe 181
Daspletosaurus 172
Daurosaurus 103
Deinocheirus xi, 4, 81–2, 84, 86, 90–2, 94–7
Deinonychus 1
Demidenko, Natalya 111
Depéret, Charles 173, 183
Deva Museum of Dacian and Roman Civilization 60
Dhouailly, Danielle 108
Diamantinasaurus 140
Diaz, Trinidad 208
digestion 160–3
Dilong 36, 132
Dilophosaurus 196
Ding, Xiaoqing 30
dinosaur

aquatic 10, 21
baby 89–90, 117, 145, 178, 183
chewing 160–1
diet 7
discovery 29, 38
diversity 3, 6, 42–4
eyes 129
gliders 6
size 5, 16, 157, 180
skin 106
skulls 16, 47, 65–79, 81, 95–6, 110, 123, 127–129, 156, 172, 177, 180–1, 197
teeth 7, 22, 129, 131, 143, 147, 158, 160, 163, 171, 177–79
Dinosaur Mailing List 103
Dinosaur Provincial Park, Alberta 67–8, 76, 81, 97, 129, 131, 175
Dinosaurs Under the Aurora (Gangloff) 112
Diplodocus 50, 67, 181
Dire Straits 170
Doda, Bajazid 49–53, 62–63
Dodson, Peter 70–71, 73, 169, 177
Doklady Earth Sciences 103
Dollo, Louis 50
Dracoraptor 214
Dracovenator 196
Dreadnoughtus 49, 155, 158
Drexel University, Philadelphia 158
dromaeosaurs 8, 28, 110
drought 170, 172, 181
Druckenmiller, Patrick 117–26, 130, 132
dwarf dinosaurs 47, 56–60, 62, 64, 129
Dyke, Gareth 50–1

Ecuador 167–8
Edmontosaurus 123–5
Edward VII, King 67
eggs 56, 84, 113, 145, 153, 155, 164–6, 175, 180, 183, 215
Egypt 5, 14, 20
  Bahariya Oasis 14–15, 17–20
  Cairo 14–15
  El Fayoum Oasis 14–15
  Gebel el Dist 15, 17
  Sahara Desert 10, 17
  Western Desert 14

234

Index

electron microscope 7, 31, 43
elephants 58, 109, 160
Elliot, David 195–6
Elsie, Robert 51–2
Ely, Ricardo 205
Endeavour, HMS 193
Eotyrannus 137
Epidendrosaurus 30–1, 40, 45
Epidexipteryx 30–1, 38, 40–1, 45
Eric the pliosaur 142
Erickson, Gregory 122–6
Erlikosaurus 84
Eromanga Sea 135, 142
Escapa, Ignacio 167
Escuillié, François 95–6
espionage 55
Europasaurus 59
Evans, Edgar 192
evolution 6, 34, 35, 37, 41, 48, 58, 60, 71, 77, 106, 107, 159, 161–2, 172, 174, 189, 202, 214
extinct 8

Federico, Fanti 213
Farke, Andy 67–70, 72–4, 76, 174, 183, 185–7
Farlow, James 73
feathered dinosaur 27, 29–30, 33–4, 40, 45, 101–109, 131
feathers 6, 35, 43
    bristles 74, 106
    colour 7, 43–4, 107–108
    dinofuzz 31, 35
    display 35
    flight 31–2, 41
    height 35–6
    insulation 35
    juvenile 44
    ornithischian 74, 100, 103–106, 159–60, 168
    pennaceous 31
    pterosaur fur 48
    ribbon-like 31, 107
    speciation 43
Ferdinand, Archduke Franz 55
Field Museum, Chicago 184, 197, 202
Fiorelli, Lucas 165

Fiorillo, Anthony 115–20, 123, 127–33
flight 7, 40–1
    gliding 31
    muscles 36
flood events 21, 78, 123, 127
Flores, Indonesia 58
Florida Everglades 19
Florida State University 122
flying squirrels 31–2
footprints 79, 119–22, 140, 144, 214
Forster, Catherine 180, 181
fossa 175
fossils
    deposition 7
    fossilisation 6, 31, 42
    black market 29, 93, 111, 113
    fake 33
    ichnofossils 140
    lacustrine 101
    opalised 141–53
    preparation 85
    private collectors 29
    prospecting 60, 82, 146, 204
    skin 101, 106, 108
    soft tissue 7, 31
    trace 19, 140, 214
Foster, Bob 148–50
four-winged dinosaurs 35–6
Fox, Richard 175
Franz, Baron Nopcsa 14, 53, 55, 58, 62–4, 191
Fricke, Henry 163
Fukuiraptor 137
Fulgurotherium 142
Futalognkosaurus 158

Gallimimus 81, 88, 97
Galman, Dave and Alan 141
Gangloff, Roland A 112
gastroliths 94
Geological Survey, US 122
Geology 116
geothermal activity 153–5, 165
Germany 14, 23, 35, 44
    Bavaria 23, 99
    Berlin 23
    Munich 15, 24
gers 83

Gestapo 25
ghost forest 197
Giganotosaurus 17, 21, 155, 162, 164, 167
ginkgoes 154, 197, 205
giraffe 160
Giza, pyramids of 58
Glacialisaurus 37, 159, 200–202
Glasgow University 162
gliding membranes 32, 40
Glossopteris 192, 194
Gobi Desert 35, 80–97, 131
    Alag Teeg 89
    Bugiin Tsav 93
    Dragon's Tomb 87
    Flaming Cliffs 83, 89
    Oyu Tolgoi copper mine 82, 84
    Polish–Mongolian expeditions 82–94
    Russian expeditions 85, 87, 89
    Tögrögiin Shiree 83
Godefroit, Pascal 95, 100–14
Godthelp, Henk 145
Golden Gate Highlands National Park, South Africa 215
Golovneva, Lina 112, 113
Gondwana 4, 10, 16, 135–8, 150, 152, 154, 164, 169, 172, 174, 178, 189, 193–4, 196, 208–10
Goodwin, Matthew 151
Google Earth 186
Gorscak, Eric 215
Goyocephale 84
Greenland 212
Grellet-Tinner, Gerald 165
Grigorescu, Dan 49, 60–1
growth rates 59, 130
Guanlong 132
Guardian 163
Gulf of Mexico 75

hadrosaurs 5, 85, 97, 106, 110, 113, 116–17, 121–5, 133, 208
Hammer, William 195–201
Hanigan, Nick and Rob 214
Hanson, Richard 195
Hanson Formation 202
Hasiotis, Stephen 115

Hatzegopteryx 48–9, 61–2
Hell Creek Formation, USA 110
Henderson, Donald 16, 77–8
Henson, Jim 1
Hernandez, Aurelio 155–6
Hews, Peter 65–6, 77–8
Hokkaido Museum 115
Hollywood 7
Holtz, Tom 97
Homalocephale 88
Homo floresiensis 58
Homo neanderthalensis 3
Homo sapiens 3
Hooker, Joseph 193
Horner, Jack 74
horns 70–8
horsetail ferns 98, 118, 123, 135
Houston Museum of Nature and Science 74
Hualianceratops 39
Huene, Friedrich von 57
Hungary 57
    see also Budapest

Ia io 34
Ibrahim, Nizar 11–14, 20–25, 23, 26
icebergs 208
Iguanodon 207
India 4, 137, 172, 185–7, 192, 209
Indian Ocean 185, 187
indricothere 160
Institute of Natural Resources, Ecology and Cryology, Chita 100, 103–105
Institute of Vertebrate Palaeontology and Palaeoanthropology, Beijing 28, 30, 38–40
Iñupiaq indigenous Alaskans 123–124, 126
Irritator 21
island dwarfing 58–9, 129
Ivantsov, Stepan 111

Japan 96
Jehol Biota 45
jewellery 213
Jianu, Coralia-Maria 60
Jiufotang Formation, China 45
Johns Hopkins University, Baltimore 54

Index

Jones, Robert 148
Jurassic, Early 37 –8, 168, 196–7, 201–202, 214–15
Jurassic, Late 6, 29, 39, 45, 98, 103
Jurassic Park 2–3
Jurassic Park III 21
Jurassic World 7

Kallokibotion 50, 61
Kazakhstan 112, 212–13
keratin 81, 106, 178
Kielan-Jaworowska, Zofia 86–8, 90–1, 96
Kikak-Tegoseak Quarry 127, 133
Knopfler, Mark 178
Kobayashi, Yoshitsugu 115
Komodo dragon 58, 158
Koppelhus, Eva 207
Korea Institute of Geoscience and Mineral Resources 93
Korea–Mongolia International Dinosaur Project 90
Kosmoceratops 72
Krause, David 173–89
Ksepka, Daniel 34
Kubacska, Andras Tasnádi 53, 55, 62–3
Kulindadromeus 36, 101, 105–10, 114
Kulindapteryx 103
Kunbarrasaurus 138, 207
Kundurosaurus 110

Lacovara, Ken 18–9, 158
Lagerstätte 99
land bridge 60
Laquintasaura 168, 208
Laramidia 75–6, 124, 129
laser fluorescence 7
Laurasia 80, 172, 196, 208
Leaellynasaura 140, 208, 210
Lee, Yuong-Nam 93
lemurs 174–5
Leptoceratops 79
Libya 20
Lightning Claw 136–7, 142, 146, 150–1
Limnosaurus 51
Liscomb, Robert 121–2
Liscomb Bone Bed, Alaska 122–5

Lita, Mustafa 53
Liverpool John Moores University 162
Loewen, Mark 69
Lost Dinosaurs of Egypt, The (Nothdurft) 14
Lü, Junchang 41
Lufengosaurus 37, 159, 201
lungfish 9, 23

Macalester College, St Paul 173
McNamara, Maria 107
Madagascar 2–4, 30, 35, 137, 170–89, 209
  Antananarivo 173, 176
  Berivotra 173, 176, 179–80, 186, 188
  Mahajanga 170, 173, 177, 184–5, 187, 188
  Morondava 187–8
Madagascar Ankizy Fund 180
Maganuco, Simone 12
Magyarosaurus 49, 56–9, 61
Mahajanga Basin Project 177, 184
Majungasaurus xi, 172–3, 177–8, 185–6
Makovicky, Peter 197–8
Malagasy 178–81, 186–7, 189
Malawi 3, 212, 215
Mali 20
mammals 14–15, 45, 79, 141, 174
mammoths 58
mangrove swamps 19
map making 54
Mapusaurus 162
marine reptiles 15, 138, 142, 159, 209
Markgraf, Richard 10, 14, 15, 18
Marsh, Othniel Charles 50, 67, 204
Martill, Dave 13, 20–1, 214
Martin, James 204
Maryańska, Teresa 87–8
Masiakasaurus 171, 178–81
mass extinction 42
Massospondylus 37, 173, 201, 215
Mauritania 20
Mayo, Alba and Oscar 156
Megalosaurus 173
Megaraptor 136
megaraptorids 136–7, 139, 150–1
melanosomes 31, 43–44, 107–109

Mexico 124, 131, 75
Microraptor 6, 28–9, 35–6, 41, 43, 106, 108
migration 120, 132, 164
Milan, Italy 12–13, 21
mining 143–52
Minmi 138
Molnar, Ralph 141, 145
Mongolia 2–3, 35, 71, 76, 80–97, 100
   Altan Ula 86, 89–90
   Nemegt 80, 82, 86, 91–2, 94, 97
   Ulaanbaatar 84–5
Mongolian Academy of Sciences 83–5
mongoose 175
Mononykus 84
monotreme 139, 142, 145, 174
Mori, Hirotsugu 123
Morocco 2, 9, 12–13, 20–1, 24–5, 18, 26, 213
   Casablanca 21
   Erfoud 11–12
   Erg Chebbi Dunes 11
   Kem Kem fossil beds 16, 20
   Sahara 11
Morrosaurus 207
Mozambique Channel 174, 177
Munich 10
Museo Argentino de Ciencias Naturales, Buenos Aires 208
Museo Paleontológico Egidio Feruglio 137, 156, 165, 167, 200
museum collection 27–8, 68
Museum of the Rockies, Montana 74, 124–5, 129, 176, 184
Museum Victoria 140
Muttaburrasaurus 144–5, 149
Myanmar 212–213

Naish, Darren 75
Namibia 170
Nanuqsaurus 5, 117, 127–8
Narmada Valley, India 186
Nasutoceratops 72
National Geographic 11, 26, 33, 96, 178
National Museum of Wales, Cardiff 214
National Science Foundation, US 195, 199, 203, 206, 209

Natural History Museum, London 1, 50, 53, 67, 107, 141
Natural History Museum of Los Angeles County 162, 198
Nature 33–4, 96–7, 142, 167, 172, 181
Nature Communications 182
Naturwissenschaften 113
Naze theropod 205
Nazis 24–5
neck length 22
Nemegtosaurus 88
Nessov, Lev 112
nesting 56, 83, 125, 155, 163, 165, 166
Neuquensaurus 59
New Britain 193
New Caledonia 193
New Guinea 193
New Perspectives on Horned Dinosaurs (Ryan) 71
New York Times 28, 196
New Zealand 193, 197, 209
Niger 16, 21, 212
nomads 83
nomina nuda 105
Nopcsa, Baron Franz 49–50, 62, 129
Nopcsa, Ilona 51
North Africa 12, 16
North America 17, 39, 71, 75, 115–34
North Korea 29
North Pole 112
Northern Hemisphere 137, 208
Nothdurft, William 14, 18
Nothofagus 193
Nothronychus 119
Nyasasaurus 69

O'Connor, Jingmai 32–3
Ohio University 215
Okavango Delta, Africa 80
Oldman River 65, 78–9
Olivero, Eduardo 194
opal, 135, 139–52
Opisthocoelicaudia 86
Orkoraptor 136
ornithischians 27, 36–7, 100–3, 105, 106–107, 160, 168
ornithomimids 4, 81, 90, 94, 97
Ornithomimus 44

238

Index

ornithopods 6, 59, 103, 112, 140, 144, 147, 207–208, 210
Orolotitan 110, 125
Osborn, Henry Fairfield 16
Osmólska, Halszka 86–8, 90, 96
Ottoman Empire 54
Owen, Richard 173, 215

pachycephalosaurs 39, 84, 124, 177
Pachycephalosaurus 39, 177
Pachyrhinosaurus 37, 71, 120, 124, 127–8, 133
Padian, Kevin 34
Padillasaurus 168
palaeobiology 64, 130
palaeobotanists 112, 203
palaeoillustrators 74
palaeontology, sexism in 184
Paleo-Arctic Research Consortium 130, 132
Pangaea 194, 196–7, 205
Paralititan 18–19, 164, 187
Parastratiosphecomyia 34
patagia 32
Permian 203, 209
Perot Museum of Nature and Science, Dallas 115, 127–133
Peru 167–168
Peyer, Bernhard 15
Pinacosaurus 89
poaching 93
Pol, Diego 137, 156–57, 164, 167, 200
Poland 88
    see also Warsaw
polar darkness 113, 117–18, 129, 135, 151, 197, 205
polar dinosaurs 4, 117–34
Polarornis 208
Polish Academy of Science 86
political instability 20
politics 24
Pompeii 35
Portugal 213
predation 162
Presley, Elvis 72
pressure receptors 22
Prince Creek Formation, Alaska 122
private collectors 29

propatagium 32
prosauropods 37, 159, 200–202
Protoceratops 75, 83, 132
Psittacosaurus 74, 75, 106, 111
Pteranodon 120
pterosaurs 32–3, 45, 47–9, 62, 120
Puerta, Pablo 156, 167

Queensland Museum 139, 141
Quetzalcoatlus 49, 61, 120

Rahonavis 171, 178–81
Randriamiaramanana, Louis Laurent 178
Rapator 136
Rapeto 181
Rapetosaurus 180–3, 187
Ratsimbaholison, Liva 186
Ravoavy, Florent 177
Raymond M Alf Museum of Paleontology, California 67, 174
Regaliceratops 5, 39, 76–8
Repenning, Charles 122
Rich, Tom 140
Ritchie, Alex 141–8
Rocky Mountains 65, 76
Romania 3
    Bucharest 51
    Hațeg 47, 49, 52, 57–62
    Oradea 59
    Szacsal 51, 53
    Transylvania 47, 56
Roniewicz, Ewa 87
Rowan University, New Jersey 18
Royal Belgian Institute of Natural Sciences, Brussels 95–6, 100
Royal Hungarian Geological Institute, Budapest 63
Royal Tyrrell Museum of Palaeontology, Canada 15, 44, 65–66
Rukwatitan 215
Russia 35, 84, 98–114
    Amur Oblast 109
    Kakanaut River 112
    Kamchatka Peninsula 112–13, 133
    Kemerovo 110–111
    Kiya River 110

239

Kulinda 100–109
Kundur 109
Lake Baikal 100–101, 110
Moscow 89, 99
Olov River 100
St Petersburg 112
Shestakovo 110–11
Siberia 2–3, 36, 83, 98, 109
Russian Academy of Sciences 100, 104
Ruxton, Graeme 162
Ryan, Michael 69, 71, 212

Saichania 84
St Mary's College of California 204
Salisbury, Steve 140, 209
salmon 128
Sanchez, Joe 78
Sander, Martin 59
sandstone 44
sandstorm 82–83
saurischians 105–106
sauropodomorphs 37, 159, 200–202
sauropods 2, 5–7, 20, 37, 48–9, 133, 143, 153–69, 180–5, 188, 201–202, 214
Saveliev, Sergei 103–105
sawfish 9
scales 102, 106, 108
scansoriopterygid 30–2, 40
Scansoriopteryx 30
Scasso, Roberto 194
Scelidosaurus 207
Science 21, 104–105, 108, 167
Scientific American 50
Sciurumimus 36
scorpions 18
Scotland 214
　　Isle of Skye 214
　　Shetland Islands 60
Scott, Captain Robert Falcon 190–4
Scott Polar Research Institute, Cambridge 194
sea levels 19
seed ferns 98
Sefapanosaurus 215
Segnosaurus 84
Serengeti 163
Sereno, Paul 16, 21

Sertich, Joseph 185–9, 206, 208
sexual dimorphism 57, 73
sexual selection 75
Shandong Tianyu Museum of Nature, Pingyi 27, 32–3, 38
Shell Oil 121–2
Siats 137
Siberian Times 111
Sibişel River 61
silica 139
Simosuchus 178
Sinitsa, Sofia 100–101, 103–105, 114
Sinoceratops 75, 132
Sinosauropteryx 34–5, 43, 106
Sinosaurus 196
sleep 161
Smith, Elizabeth 139, 140–51, 184
Smith, Josh 18, 23
Smith, Nathan 198, 200–204, 209–210
Smith Woodward, Arthur 53, 68
Smith Woodward, Lady 53
snakes 15, 18, 155, 168, 213
Society of Vertebrate Paleontology, US 102, 185
South Africa 3, 37, 159, 173, 192, 196, 201–202, 210, 215
South America 2, 4, 49, 59, 137, 153–99, 172, 178–9, 185, 189, 193, 207–209, 211, 215
South Dakota School of Mines 204
South Korea 96, 165
southern beeches 193–4
Southern Hemisphere 172–3, 177, 189, 196, 208–209
Spain 213
species recognition 73
spiders 18
Spinops 5, 39, 69
Spinosaurus 3, 10–13, 15–16, 20–6
Star Wars 80
State University of New York 180
Sternberg, Charles and Levi 67–9
sternum 36
Steropodon 142
Stony Brook University, New York 173, 176, 185, 189
Stromer, Ernst 10, 14, 17–18, 24–5, 191

240

Index

Stromer, Gerhart 25
Stromer, Ulman 25
Stromer-Baumbauer, Rotraut von 24
Stromer's three-theropod riddle 17, 20, 23
Struthiosaurus 51
Styracosaurus 39, 68–9, 71–2
Suchomimus 21
Sues, Hans-Dieter 19, 213
Suess, Eduard 51, 64
Sullivan, Corwin 31–33, 40–1, 45
Switek, Brian 179

Tachiraptor 168
taiga forest 116
Tanke, Darren 65, 67, 69, 73, 76
Tanzania 215
Tarbosaurus 17, 81–2, 87, 92–4
Taylor, Michael 159–60
Telmatosaurus 49, 51, 56, 58, 61
tenrecs 174, 175
Terra Nova expedition 192
Tethys Sea 59
Texas Christian University 195
therizinosaurs 36, 84, 94, 119, 131
theropod 6, 16, 32
    forelimbs 87, 90, 172
Therrien, François 16, 44
Tianyulong 27, 106
Tianyuraptor 27
Tiaojishan Formation, China 45
Tibetan Plateau 83
Tierra del Fuego 193
Timurlengia 213
Timimus 140
titanosaurs 2, 49, 111, 153–69
Titanosaurus 57
Tomsk State University, Russia 110–11
Triassic 69, 108, 159, 203
Triceratops 2, 39, 65–6, 70–5, 77, 110–11, 160
Trinisaura 208
Troodon 128–29
troodontids 110, 128–29
Tsagantegia 84
Tsedevedamba, Oyungerel 96
Tsogtbaatar, Khishigjav 83, 85, 96
tughrik 93, 95

Tunisia 19–20, 212–13
Turkey 55
Turkmenistan 212
turtles 15, 19, 50, 57, 61, 79, 110, 138–39, 145–6, 151, 155
Tykoski, Ronald 128
tyrannosaurs 5, 8, 36,112, 124, 132, 213
Tyrannosaurus rex xi, 4, 8, 10, 15–17, 21, 67, 72, 80, 82, 84, 90, 136, 151, 154, 158, 164, 172, 213
Tyrannotitan 155, 158

Ugrunaaluk 5, 117, 123–5
Umoonasaurus 142
United Kingdom 21, 109
    London 53
United States 16, 110
    Chicago 21
    Utah 163
    Wyoming 66–7, 163
Université d'Antananarivo 177, 181, 186
Université Hassan II, Casablanca 12
Université Joseph Fourier, Grenoble 108
University of Alaska Museum of the North 117, 121–2
University of Bologna 213
University of Bonn 59
University of Bristol 49, 104, 107, 159
University of Bucharest 60
University of Calgary 44
University of California, Berkeley 122
University of Chicago 16
University College Cork 107
University of Edinburgh 96, 107
University of Kansas 115
University of Lyon 173
University of Manitoba 175
University of Maryland 97
University of New England, Australia 89, 137, 149
University of New South Wales 145
University of Pennsylvania 18, 71
University of Portsmouth 13
University of Queensland
University of Tübingen 57

University of Witwatersrand,
    Johannesburg 215
Uzbekistan 3, 212–213

vacuum cleaner 162
Vagaceratops 69
Vahiny 183
Vegavis 208
Velociraptor 1, 6–7, 41, 61, 83–4, 129, 205
Venezuela 3, 168
Vienna Academy of Sciences 52
Vienna Museum of Natural History 1
Vince, Gaia 42
Vintana 189
volcanic ash 35, 101, 109, 155–6, 165
volcanic sediments 44
vulcanism 98
vultures 108

Walgett Spectator 144
Wang, Jianrong 29
Wang, Tao 38
Wang, Xiaoli 28
warm-blooded 4, 49, 113, 130
Warsaw 86
Wedel, Mathew 159–160
Wegener, Alfred 194
Weishampel, David 51, 54, 60, 63–4
Wendiceratops 72
whales 22, 118, 194
White Cliffs of Dover, England 56
wildebeest 163
Wilkinson, David 162

Willis, Paul 142
Wilson, Edward A 192
Wilson, Jeff 183
Wired 48
Wolff, Ewan 73
wolves 126
Wopenka, Joe 144
World Dinosaur Valley, Lufeng 37–8
World War I 15, 47, 55, 63, 67
World War II 10, 87
Wrangel Island, Arctic 58

Xenoceratops 69
Xenoposeidon 69
Xiaotingia 28, 45
Xing, Lida 213
Xing, Xu 6, 28–30, 32, 38, 40, 43, 132

Yanliao Biota 45
Yi qi 6, 34, 36, 40–1, 44–5
Yinlong 39, 75, 132
Yixian Formation, China 41, 45, 99, 109
Yucatan Peninsula, Mexico 113
yurts 83
Yutyrannus 8, 29, 36, 38, 106, 172

Zalmoxes 56, 58, 61
Zelenitsky, Darla 44
Zhenyuanlong 8, 41
Zheng, Xiaoting 30, 32–33
Zimbabwe 184
Zouhri, Samir 12–13, 20–1